A Decision-Makers Guide to Public Private Partnerships in Airports

Airport development is critical to economic growth and poverty reduction. This book will help decision-makers assess whether Public Private Partnerships (PPP) might be a viable option to meet their airport development requirements. It walks the reader through the airport PPP process, from early preparation to bringing the project to market and managing the project during implementation. The book will help eradicate misconceptions about the role of the private sector in airport infrastructure.

A Decision-Makers Guide to Public Private Partnerships in Airports provides an essential guide for those in a position to make decisions linked to airport development, to their advisers, their staff and also to students wishing to understand airport PPP.

Andy Ricover is an air transport economist and has been advising governments on aviation economics, policy formulation, regulation, valuation and ownership reforms since 1994. He has worked in over 240 air transport projects in 92 countries for major consultancy firms, operators, bid consortiums, governments, law firms and is a regular consultant to the major development banks and agencies (World Bank Group, IDB, CAF, AfDB, EBRD, USAID, etc.). He has advised on numerous PPP processes (for airports and airlines) for the public and the private sectors from design (institutional, strategic and regulatory) through implementation (due diligence, tendering, financing). As an expert witness, Andy has provided written and oral testimony in over a dozen arbitration cases concerning airlines and airports. He is a regular speaker at conferences and is an active lecturer at universities and airport authorities.

Jeffrey Delmon has advised governments, sponsors and lenders on infrastructure finance and PPP since 1994, and has been with the World Bank since 2005. He has advised governments on airport PPP in Europe, Africa and East Asia; he was strategic adviser for the government in Saint Petersburg, Russia, for the €1.2bn Pulkovo Airport which reached financial close in 2010. Prior to joining the World Bank, Jeff spent 11 years in Paris and London advising on infrastructure and PPP projects globally at the law firms of Allen & Overy and Freshfields. He is author of numerous books, articles and blogs on PPP, and his most recent books include: *Public Private Partnership Projects in Infrastructure: A practical guide for policy makers* (second edition, 2017), and *Private Sector Investment in Infrastructure: Project finance, PPP projects and PPP programs* (third edition, 2016). He is adjunct associate professor on PPP at the National University of Singapore.

A Decision-Makers Guide to Public Private Partnerships in Airports

Andy Ricover and Jeffrey Delmon

LONDON AND NEW YORK

First published 2020
by Routledge
2 Park Square, Milton Park, Abingdon, Oxon OX14 4RN

and by Routledge
52 Vanderbilt Avenue, New York, NY 10017

Routledge is an imprint of the Taylor & Francis Group, an informa business

© 2020 Andy Ricover and Jeffrey Delmon

The right of Andy Ricover and Jeffrey Delmon to be identified as authors of this work has been asserted by them in accordance with sections 77 and 78 of the Copyright, Designs and Patents Act 1988.

All rights reserved. No part of this book may be reprinted or reproduced or utilised in any form or by any electronic, mechanical, or other means, now known or hereafter invented, including photocopying and recording, or in any information storage or retrieval system, without permission in writing from the publishers.

Trademark notice: Product or corporate names may be trademarks or registered trademarks, and are used only for identification and explanation without intent to infringe.

British Library Cataloguing-in-Publication Data
A catalogue record for this book is available from the British Library

Library of Congress Cataloging-in-Publication Data
Names: Ricover, Andy, author. | Delmon, Jeffrey, author.
Title: A decision-makers guide to public private partnerships
in airports / Andy Ricover and Jeffrey Delmon.
Description: Milton Park, Abingdon, Oxon ; New York, NY : Routledge, 2020. |
Includes bibliographical references and index.
Identifiers: LCCN 2019036599 | ISBN 9780367266783 (hardback) |
ISBN 9780429294556 (ebook)
Subjects: LCSH: Airports–Economic aspects. |
Public-private sector cooperation.
Classification: LCC HE9797.4.E3 R53 2020 | DDC 387.7/360684–dc23
LC record available at https://lccn.loc.gov/2019036599

ISBN: 978-0-367-26678-3 (hbk)
ISBN: 978-0-429-29455-6 (ebk)

Typeset in Goudy
by Newgen Publishing UK

All images © Andy Ricover

 Printed in the United Kingdom
by Henry Ling Limited

Contents

Preface		vii
Acknowledgements		xi
Glossary		xii
1	Introduction	1
2	World trends in air transport	15
3	The business of airports	25
4	Airport planning	57
5	Air transport regulation	67
6	Airport PPPs around the world	100
7	Financing airport PPP	139
8	Airport PPP preparation	152
9	Procuring airport PPP	175
10	PPP agreements	190
11	Implementation, monitoring and evaluation	219
	Bibliography	241
	Index	247

Preface

> Writing is nothing more than a guided dream.
> Jorge Luis Borges, Argentinian writer (1899–1986),
> Preface to *Dr. Brodie's Report [El informe de Brodie]* (1970)

Bigger, better, faster, cheaper – the pressure on government is massive: to deliver infrastructure, to create jobs, to provide new economic opportunities. Airports are complex pieces of infrastructure: critical for economic development, jobs and wealth creation, yet perceived as an asset for the upper and middle classes alone, and thus creating nervousness about spending too much public capital on airport development. While publicly delivered airport services are globally less efficient and impactful than where private partners are involved, private participation makes people nervous. The spectre of privatisation persists. There is concern that private investors will take advantage of less sophisticated public agencies. Hence the need for this book: to debunk some of the myths around airport PPP, to explore the challenges, and to share lessons learned from the global airport PPP market.

This book describes the key issues and steps that are needed to prepare an airport PPP project, whether international hub, gateway or small domestic airport. It is focused on airport PPP and therefore does not cover in detail issues related to aviation access policy or technical issues in airport infrastructure, where literature abounds.

Different PPP structures are used to deliver services based on demand. In many countries, the whole airport is offered for private sector operation through a concession model. Airport PPP is not easy. Successful PPP projects require careful preparation and experienced advisors during feasibility and transaction implementation. More importantly, national and regional governments must understand the issues specific to PPP in airports and bridge the gap between public and private objectives. This book therefore focuses on issues specific to airport PPP. For a general discussion of issues relevant to PPP, there are numerous texts and articles on the topic. The reader may wish to consider:

- The PPP Legal Resource Center (https://ppp.worldbank.org/public-private-partnership/about-ppplrc-ppp-legal-resource-center)

- Jeffrey Delmon, *Public Private Partnerships in Infrastructure: An Essential Guide for Policymakers* (2017)
- Jeffrey Delmon, *Private Sector Investment in Infrastructure: Project Finance, PPP Projects and PPP Programs* (2016)

Airports and development

Airports are key economic instruments. They can generate economic growth, create jobs, create opportunities, and reduce poverty. The economic context of airports will be explored, considering the role of airports, the opportunities they provide and the economic space in which they are found. While PPPs are in many ways a commercial construct, airports must always contemplate their important economic context in the country, and the region.

Airport business

Airports are a dynamic piece of core infrastructure, and at the same time they are market-driven and market drivers. They are a public service and commercial generator. This book will explore the business of airports, how they generate revenues, different airlines, different commercial activities, how airports link with markets and market-makers. It explores demand, looks at the cost base of running an airport, where revenues come from and how to maximise them. It will also explore the cost of running an airport and the challenges that cause those costs to rise. It will consider the financial context of PPP, what represents value for money, and how a normal PPP would manage questions of costs and revenues (for example, fees and charges, and incentives). It maps out potential cost reductions and revenue increases available through PPP (airside/landside, aeronautical/non-aeronautical), the type of government support that can be used to fill any revenue gaps, and how the government might share in revenues that exceed project requirements. An airport PPP will rely on demand forecasts (which are difficult to get right), how investors and lenders perceive demand risk, and how a policy maker can help assuage this perception of risk. The chapter will also explore sources of financing.

Airport planning

An airport needs to fit into a local, national and regional planning context, to ensure it delivers the facilities and services needed (given the demand and nature of aircraft, passengers and cargo likely to use the facility) as well as to comply with safety and security standards. This chapter discusses airport planning, from the Master Plan that maps out the development of airports and aviation over the medium to long term in the city and the country in question, to the strategic planning that drives the design and development of an individual airport. The investment strategy drills down into the detail of an airport, including size of aircraft, size and number of terminals, technology, traffic, public versus private management and transport links to nearby cities. Airport facilities may use the presence of the airport to deliver other services, for example, transport links to encourage industrial, technological and commercial activities in and around the airport.

Airport regulation

As a public transport facility that delivers critical economic services, airports are generally regulated, for safety and security, for economic context such as fees and charges, and for linkages to other transport functions in the city, the country and the region. Airports are also subject to international structures that create a regulatory environment, such as ICAO standards and recommended practices, IATA levels of service, bilateral arrangements (including open skies) and regional arrangements designed to encourage regional connectivity. These regulatory functions are meant to enable airports to deliver their services better and in a more structured environment, but require careful design to ensure that PPP arrangements fit with regulatory context.

The fundamentals of airport PPP

This book presents the basics of airport PPP, the actors involved, the scope of the PPP, labour and other implications of PPP, different forms of PPP and how to select the right PPP model. It will also start discussion on the process of finding the right partner, as discussed and mapped out below.

Airport PPP preparation

The preparation process of an airport PPP involves a series of steps, each building on the last. It starts with the selection of a project likely to be successful as a PPP. Once the project is selected, a project team must be appointed, with the right mixture of skills and seniority. The project team will oversee the development of a pre-feasibility study, to do a preliminary assessment of the project. If successful, a feasibility study is next, testing all aspects of the project, identifying the best model of PPP to use and preparing for the procurement process. This will include market sounding, stakeholder consultation and internal approvals. The feasibility study is generally implemented by specialist transaction advisers, to make sure the assessment is somewhat objective and technically rigorous. The government will use this feasibility study to identify the best PPP model for the project, develop the right financial structure and test it with the market.

Procuring airport PPP

After successful completion of a feasibility study and obtaining relevant approvals, the team will move to the procurement phase, using a competitive process to get the best private investors and the best deal available. This will start with pre-qualification of a short list of high-class investors, then on to bidding, when the short list competes to provide the most responsive technical proposal and the most advantageous financial proposal. The process will include criteria used to test the technical and financial bids. Bidding is followed by awarding the contract to the most responsive bidder and financial close, further to engagements with the financiers.

PPP agreements

The project will be established through contractual arrangements, anchored on a PPP Agreement. This document will set out the agreement between the parties on the project, rights, obligations, fees, penalties, liabilities and the like. It will set out timing and arrangements for mandatory investments, safety and security standards, required levels of service, access to land, construction process, performance indicators, penalties, government support, government revenue share and dispute resolution. It will describe required insurances, and how specific risks will be allocated between the parties, including *force majeure*, relief events and termination. It will also contemplate financing and security arrangements, where appropriate.

Implementation, monitoring and evaluation

Once the bidding is done and financial close achieved, the project needs to be implemented. For government, this means creating an implementation plan, including operation guidelines, construction monitoring, fees and charges regime, options for selling down equity, refinancing, dispute management and renegotiation. The implementation stage is critical to the success of the project, but is often poorly prepared or resourced by government.

Acknowledgements

The authors would like to thank Vickram Cuttaree, Gevorg Sargsyan, Alex Webber, Angelica Botero, Federico Masri, Lincoln Flor and the friends and colleagues who have provided critical advice, review and guidance on this text. In particular, we would like to thank Ellis Juan and Martin Sgut for their leadership and guidance over the years. And most importantly, the authors would like to thank San, Liat, and Mei; Vicky, Alex and Natasha – for their support and patience, long may it last!

This book builds on an unpublished study by Andy Ricover, Vickram Cuttaree and Jeffrey Delmon titled "Airport Development through Public-Private Partnerships: Guide for decision makers".

The findings, interpretations, and conclusions expressed herein are those of the authors and should not be attributed in any manner to the World Bank or the International Finance Corporation, their affiliated organisations, or to the members of their Board of Executive Directors or the countries they represent. This text does not constitute legal advice, and does not substitute for obtaining competent legal counsel (readers are advised to seek the same) when addressing any of the issues discussed in this text.

Glossary

This glossary of terms, abbreviations and acronyms is included exclusively for use as a reference aid and therefore should not be considered an exhaustive or complete discussion of any of the terms set out below, or indeed of all the terms relevant to PPP or project finance. Definitions are generally given under their spelt-out form, and the abbreviation refers to the spelt-out form.

ACI	Airports Council International
ADEMA	Aéroports de Madagascar
ADRM	Airport Development Reference Manual (IATA)
AENA	Spanish airports company
AERA	Indian Airports Economic Regulatory Authority
aeronautical revenues	revenues from aeronautical fees and charges for the provision of airport services to users (airlines or aircraft operators, passengers and cargo owners)
AHM	Airport Handling Manual (IATA)
airlines	a company that carries passengers and/or cargo and collects revenues. For the purpose of this book, airlines are also aircraft operators of any other type, including not on revenue purposes
Air Traffic Movements (ATM)	a landing or a take-off of an aircraft from an airport
ANA	Portuguese airports company
ANSP	air navigation service provider
APAC	Asia Pacific
ASA	Air Service Agreement
ASEAN	Multilateral Agreement on Air Services of the Association of Southeast Asian Nations
AVSEC	aviation security (safeguarding international civil aviation against acts of unlawful interference)
BALIs	Chilean tender conditions
BASA	Bilateral Air Service Agreement
BID	Inter-American Development Bank
BIT	Bilateral Investment Treaty

Glossary   xiii

BOOT	Build Own Operate Transfer
BOT	Build Operate and Transfer
BSP	Billing and Settlement Plan (IATA)
CAA	Civil Aviation Authority
CAN	Comunidad Andina (Andean Community trade bloc)
capitalised interest	accrued interest (and margin) which is not paid but added ("rolled up") to the principal amount lent at the end of an interest period; see, for example, interest during construction
CARICOM	Caribbean Community economic organisation
concession agreement	the agreement with a government body that entitles a private entity to undertake an otherwise public service
conditions precedent (CPs)	conditions which must be satisfied before a right or obligation accrues, i.e. the matters which have to be dealt with before a borrower will be allowed to borrow under a facility agreement (these will be listed in the agreement)
construction contract	the contract between the project company and the construction contractor for the design, construction and commissioning of the works
contracting authority	the party which grants a concession, a license or some other right
CPI	Consumer Price Index
credit risk	the risk that a counterparty to a financial transaction will fail to perform according to the terms and conditions of the contract (default), either because of bankruptcy or any other reason, thus causing the asset holder to suffer a financial loss; sometimes known as default risk
CUTE	Common Use Terminal Equipment
debt-to-equity (D/E) ratio	the proportion of debt to equity, often expressed as a percentage (the higher this ratio, the greater the financial leverage of the firm); also known as gearing
debt service	payments of principal and interest on a loan
debt service cover ratio (DSCR)	the ratio of income to debt service requirements for a period; also known as the cover ratio
debt service reserve	an amount set aside either before completion or during the early operation period for debt-servicing where insufficient revenue is achieved
defects liability period	the period during which the construction contractor is liable for defects after completion
direct agreement	an agreement made in parallel with one of the main project documents, often with the lenders or

xiv *Glossary*

	the contracting authority; step-in rights and other lender rights are often reinforced or established through direct agreements between the lenders and the project participants
discount rate	the rate used to discount future cash flows to their present values, often based on a firm's weighted average cost of capital (after tax) or the rate which the capital needed for the project could return if invested in an alternative venture (a higher discount rate may be used to adjust for risk or other factors)
DME, VOR and NDB	different types of radio navigational beacons that provide assistance for position and/or distance in relation to the aircraft
DRB	dispute resolution boards
EBITDA	earnings before interest, taxes, depreciation, and amortisation
ECA	export credit agency
ECAA	EU Internal Market and the European Common Aviation Area
economic rate of return, also economic internal rate of return (EIRR)	the project's internal rate of return after taking into account externalities (such as economic, social and environmental costs and benefits) not included in financial IRR calculations
E&F	Enhancement and Financing Services (IATA)
EHCAAN	Egyptian Holding Company for Airport & Air Navigation
environmental impact assessment (EIA)	an assessment of the potential impact of a project on the environment; results in an environmental impact statement
environmental impact statement (EIS)	a statement of the potential impact of a project on the environment; the result of an environmental impact assessment which may have been subject to public comment;.
EPC contract	engineering, procurement and construction contract (i.e. turnkey construction contract)
equity	the cash or assets contributed by the sponsors in a project financing; a company's paid-up share capital and other shareholders' funds; for accounting purposes, it is the net worth or total assets minus liabilities
ET	Enterprise Team
EUR	European Union euro
FAA	US Federal Aviation Administration

Glossary xv

financial close	in financing, the point at which the documentation has been executed and conditions precedent have been satisfied or waived; drawdowns become permissible after this point.
financial internal rate of return (FIRR)	see **internal rate of return**
fiscal space	capacity in a government's budget (including borrowing capacity) that allows it to provide or access resources for a desired purpose without jeopardising the sustainability of its financial position or the stability of the economy or otherwise breaching restrictions created by its own national laws or by supranational bodies or by lenders (in particular large lenders such as the IMF or World Bank)
force majeure	events outside the control of the parties and which prevent one or both of the parties from performing their contractual obligations
greenfield	often used to refer to a planned facility which must be built from scratch, without existing infrastructure
hub bank	the period of time in a day when flights arrive at the airport and passengers change flights for outward connections
IAI	Israel Aircraft Industries
IATA	International Air Transport Association
ICAO	International Civil Aviation Organization
ICC	International Chamber of Commerce
ICSID	International Centre for Settlement of Investment Disputes
IFC	International Finance Corporation
IFI	International Financial Institution
IGOM	*Ground Operations Manual* (IATA)
Instrument Landing System (ILS)	a system of antennas located at the airport (and its vicinity) which provides radio-navigational assistance to pilots for a precision approach and landing under reduced visibility conditions
intercreditor agreement	an agreement between lenders as to the rights of different creditors in the event of default, covering such topics as collateral, waiver, security and set-offs
interest during construction (IDC)	interest accumulated during construction, before the project has a revenue stream to pay debt service, is usually rolled up and treated as capitalised interest
internal rate of return (IRR)	the discount rate that equates the present value of a future stream of payments to the initial investment; see also **EIRR**

KPI	key performance indicators
LCC	Low Cost Carrier
LCIA	London Court of International Arbitration
limited recourse debt	see **non-recourse**
liquidated damages (LDs)	a fixed periodic amount payable as a sanction for delays or substandard performance under a contract; also known as a penalty clause
LoS	Level of Service (IATA)
MASA	Multilateral Air Service Agreement
MALIAT	Multilateral Agreement on the Liberalization of International Air Transportation
maximum take-off weight (MTOW)	the maximum weight of an aircraft at which it can take off
mezzanine financing	a mixture of financing instruments, with characteristics of both debt and equity, providing further debt contributions through higher-risk, higher-return instruments; sometimes treated as equity
MLA	multilateral agency, for example the World Bank
MLIT	Japan Ministry of Land, Infrastructure, Transport and Tourism
MOP	Chilean Ministry of Public Works
MTR	minimum technical requirements
NAP	Canadian National Aviation Policy
NAS	Canadian National Airports System
NDB, VOR and DME	different types of radio navigational beacons that provide assistance for position and/or distance in relation to the aircraft
net present value (NPV)	the discounted value of an investment's cash inflows minus the discounted value of its cash outflows; to be adequately profitable, an investment should have a net present value greater than zero
non-recourse (limited recourse)	the lenders rely on the project's cash flows and collateral security over the project as the only means to repay debt service, and therefore the lenders do not have recourse to other sources, for example shareholder assets; more often, non-recourse debt is actually limited recourse debt
off-balance sheet liabilities	corporate obligations which do not need to appear as liabilities on a balance sheet, e.g. lease obligations, project finance and take-or-pay contracts
OPIC	Overseas Private Investment Corporation
ORAT	Operational Readiness and Airport Transfer
ORSNA	Argentinean body for regulation of the national airports system

Glossary xvii

OSITRAN	Peruvian regulatory agency for supervision of investments in transport infrastructure
Passenger Facility Charge (PFC), also Passenger Service Charge (PSC)	charge levied on passengers for the use of the terminal building and related infrastructure
PDF	Project Development Fund
performance bond	a bond payable if a service is not performed as specified; some performance bonds require satisfactory completion of the contract while others provide for payment of a sum of money for failure of the contractor to perform under a contract
PPH	peak profile hour
PPP	Public Private Partnership
pre-qualification	the process whereby the number of qualified bidders is limited by reviewing each bidder's qualifications against a set of criteria, generally involving experience in the relevant field, capitalisation, site country experience, identity of local partners and international reputation
project financing	a loan structure that relies for its repayment primarily on the project's cash flow, with the project's assets, rights and interests held as secondary security or collateral; see also **limited recourse and non-recourse financing**.
PSC	Public Sector Comparator
RAB	regulatory asset base
refinancing	repaying existing debt by obtaining a new loan, typically to meet some corporate objective such as the lengthening of maturity or lowering the interest rate
RESA	Runway End Safety Area
reserve account	a separate amount of cash or a letter of credit to service a payment requirement such as debt service or maintenance
return on assets (ROA)	net profits after taxes divided by assets; this ratio helps a firm determine how effectively it generates profits from available assets
return on equity (ROE)	net profits after taxes divided by equity investment
return on investment (ROI)	net profits after taxes divided by investment
SARPs	Standards and Recommended Practices
security	a legal right of access to value through mortgages, contracts, cash accounts, guarantees, insurances,

	pledges or cash flow, including licences, concessions and other assets; a negotiable certificate evidencing a debt or equity obligation/shareholding
shareholders' agreement	the agreement entered into by the shareholders of the project company which governs their relationship and their collective approach to the project
SMP	substantial market power
SOE	state-owned enterprise
special purpose vehicle	an entity created to undertake a project in order to protect the shareholders with limited liability and limited or non-recourse financing
sponsor	a party wishing to develop/undertake a project, a developer, or a party providing financial support
STC	(Mexico) Secretary of Transports and Communications
step-in rights	the right of a third party to "step in" to the place of one contractual party where that party fails in its obligations under the contract and the other party to the contract has the right to terminate the contract
subordinated debt	debt which, by agreement or legal structure, is subordinated to other (senior) debt, allowing those (senior) lenders to have priority in access to amounts paid to the lenders by the borrower from time to time, and to borrower assets or revenues in the event of default; this priority may be binding on liquidators or administrators of the borrower; it does not include reserve accounts or deferred credits
tranche	a separate portion of a credit facility, perhaps with different lenders, margins, currencies and/or term
TSA	Technical Service Agreement; also US Transportation Security Administration
turnkey construction	the design and construction of works to completion, so that they are ready to produce cash flow
ULDR	Unit Load Devices Regulations (IATA)
ultra vires	an act outside the scope of one's authority
UNCITRAL	United Nations Commission on International Trade Law; responsible for the Convention on the Recognition and Enforcement of Foreign Arbitral Awards amongst others
USOAP	Universal Safety Oversight Audit Programme
USD	United States dollars
VAT	value added tax
VfM	value for money

VOR, NDB and DME	different types of radio navigational beacons that provide assistance for position and/or distance in relation to the aircraft
weighted average cost of capital (WACC)	the total return required by both debt and equity investors expressed as a real post-tax percentage on fund usage
WLU	work load unit
works	a technical term in construction identifying the entirety of the facilities and services to be provided by the construction contractor
WSG	IATA worldwide slot guidelines

1 Introduction

The better is the enemy of the good. *[« Le mieux est l'ennemi du bien. »]*
Voltaire (1694–1778), French writer, deist and philosopher,
The Prude [La Bégueule] (1772)

Despite their critical contribution to economic growth, development, job creation, trade and mobility, airports in many countries do not meet international safety and security standards, or having exceeded their capacity are unable to meet demand. As a result, safety and security standards decline, together with service levels, which constrains economic development and employment. In order to comply with international standards and meet the demand for air transportation, national and regional government administrations face pressure to improve operations, invest in facilities, increase airport capacity and upgrade services. Often, massive investment is required for airport upgrades, which has led many countries to consider private sector investment and operation.

The public sector provides financing for the vast majority of infrastructure services. However, scarce public resources and limited fiscal space (the credit capacity/right to borrow money) means that infrastructure faces stiff competition from alternative uses of public funds. Airports face an even greater battle for attention from policy makers as they are often, incorrectly, viewed as a service only for the rich.

Public private partnerships (PPP) represent an approach to procuring infrastructure services that is radically different to traditional public procurement. It moves beyond the client–supplier relationship where government hires private companies to supply assets or a service. PPP is a partnership between public and private to achieve a solution, to deliver an infrastructure service over the long term. It combines the strength of the public sector's mandate to deliver services and its role as regulator and coordinator of public functions with the private sector's focus on profitability and commercial efficiency. PPP is ultimately flexible, limited only by the creativity of those involved and their access to funding. Chapter 6 discusses PPP in the context of airports and the models most commonly used.

This chapter introduces the topics critical to decision makers when contemplating airport PPP, looking first at common myths associated with airport PPP and then reviewing key issues and risks.

1.1 Mythbusters[1]

Myths surrounding airport PPPs often distract policy makers from the opportunities that these transactions can offer. But an open mind, commercial awareness, and the use of experienced advisers can cut through the clamour.

"PPP is free money"

This is often part of the sales pitch of a consultant looking for work, or a company hoping for a quick contract. But an airport PPP is not a free lunch. There may be a belief in government that the private sector will deliver the infrastructure with no engagement or involvement by the government. This is clearly not true. Airports require commitment and often serious public funding from government. Challenges will arise, and government needs to be engaged in the project to help resolve those issues.

---

1 Adapted from Ricover and Delmon, "Mythbusters" (2016).

"Don't bother with bidding, just select a good operator and go; it is much faster and cheaper"

A company or foreign government shows up in government offices with promises of instant gratification, a simple, easy solution to all of their problems. This generally results in years of discussions, negotiations, signed memorandums of understanding, ribbon cutting, but no progress, only delays, costs and frustration, for everyone involved; or else the investment is delivered, but at a much higher cost and with less attractive terms. And you can understand the private sector, it wants a deal and is frustrated by the slow pace of government efforts, yet golden opportunities lie, seemingly, just out of reach. Why not try to push, even if procurement rules in most countries require a more thorough approach to procurement?

The easy way does not work, if it means trying to short-cut good preparation and robust competition. Government needs to take time to work out what it wants, when it wants it, what if anything it is willing to pay or guarantee, and how different project risks are going to be managed/allocated. Once the government knows its project and its role, a competitive process should be used to select the private partner. Not because it is impossible to choose a good company directly, or to find bench-marks to compare pricing and terms; but rather because competition helps to get the best deal, helps to demonstrate that the project is awarded properly and cleanly and to give opportunities to the best investors. Yes, it takes time; yes, it requires funding and effort and good staff, but the results are worth the investment, and the "faster" non-competitive approach usually takes longer.

Ain't no such thing as a free lunch. If it seems too good to be true, it probably is.

"The private sector is always more efficient than the public sector"

While the private sector has the potential to deliver efficiently, with the right incentives embedded in the PPP agreement and a competitive process that motivates best technology and methodologies, it may not. In some cases the relationship between public and private breaks down, resulting in loss of trust, underinvestment and disputes. The failure is often linked to the design of the PPP, a failure of one or both sides to do thorough due diligence, or an effort to misuse the PPP for purposes other than the best interests of the airport. Often, the break-down is linked to a change in Government, where the new Government raises questions about deals done by previous Governments. Examples include Terminal 3 in Manila with Fraport, Terminal 2 in Budapest with ADC & ADMC, Margarita Porlamar in Venezuela with Zurich Flughafen, Male in the Maldives with GMR/Malaysia, and Bucharest Henri Coandă with EDF Services.

"PPP is all about getting more financing"

PPP is not about maximising private capital; it is about finding the most efficient and effective combination of public and private. It is a partnership. There needs to be enough private commitment to keep them engaged, but more is not necessarily better. Rather, there should be the right amount of private funding, in the right balance of debt, equity and other financial instruments; the right amount of public financing, in order to achieve a robust and successful project, something sustainable, that will survive change, challenges and chaos.

"PPP financing is cheaper"

Generally speaking, a sovereign government will be able to obtain financing at a lower cost than the sponsors or the project company.[2] However, public financing is often less efficient due to public sector incentive mechanisms being focused more on political priorities than on cost and time efficiency, and lack of institutional capacity to control costs, time for completion, changes and other risks that result in higher construction and operation costs. Private sector financing may therefore prove – in certain circumstances – less expensive, less time-consuming and more flexible to arrange or more practical than public sector financing. The private sector can provide new sources of finance (in particular where fiscal space or other constraints limit availability of government financing), include clear efficiency incentives on the project, invigorate local financial markets and manage project risks in a more efficient manner. But not always. The examples of poor performance by private operators exist. The PPP agreement needs to create the right incentives, and the government needs to stay involved, to ensure that the project goes well. This is not an easy solution, and the government needs to ensure that PPP provides better value for the country than would other solutions.

"Airport PPP is only for rich countries"

It is true that the developed world uses PPP extensively. In 1966, the ownership and control of Heathrow, Gatwick, Stansted and Prestwick airports were transferred to the British Airports Authority, based on the Airport Authority Act 1965. This corporatisation was a first step toward privatisation of UK airports. Corporatisation can help improve sector governance, by creating a clear separation between airport authorities and government regulators. This improved accountability and corporate governance through a board structure allows better identification of areas for improvement and financial viability of an airport. Corporatisation can also help in responsiveness to customer needs, by creating a

---

2 Lower interest rates obtained by a government reflect the contingent liability borne by taxpayers (see Klein's working paper for the World Bank, "Risk, Taxpayers and the role of government in project finance" (1996)). Thus, the risk that results in higher private finance interest rates reflects actual project risk and is subsidised by taxpayers to achieve the lower public finance interest rates. Since the private sector is best placed to manage most of the commercial risk in infrastructure projects, it is argued that private finance is the most efficient method of financing infrastructure; the inherent subsidy of public finance is more appropriately used in other areas.

closer link between the customer and the entity operating the airport. In 1987, the airport assets were transferred to a public liability company, BAA plc, and shares were offered for sale on the London Stock Exchange. New Zealand's largest airports were corporatised in the late 1980s with the government gradually selling down its shareholding.[3]

In the US airport industry, early privatisation was developed under the FAA statutorily authorised Airport Privatization Pilot Program. In the US, airport PPPs generally target projects requiring significant capital investment, for example the terminal projects at JFK and LaGuardia, the Great Hall project in Denver and the Paine field development in Seattle.

Fully private airport development in the US is relatively limited. Branson airport, in Missouri, is privately owned and developed, as a commercial passenger airport. It had difficulty mobilising finance and has defaulted several times on its debt obligations.[4]

Australia's airport privatisation programme established long leases for Brisbane, Melbourne and Perth airports in 1997. Sydney followed in 2002.

However, airport PPPs are also common throughout the developing world. Probably the first airport to be entirely developed and operated by the private sector, under the form of a full privatisation, is Punta Cana International Airport, in the Dominican Republic. In fact, the greatest concentration of airport concession contracts is in Latin America, with over 100 airports having been improved, expanded and operated by private investors. The concession model has also been implemented in other regions like South Asia and Eastern Europe.

In 2008, Saudi Arabia awarded management contracts to Changi airport group and Fraport. In 2009, a 25-year concession was awarded for Medina under a consortium led by TAV. Jordan launched the concession for Queen Alia airport in 2007. Egypt issued concessions for two regional airports and Cairo International airport is operated under a management contract with Fraport. Tunisia has two airport concessions operated by TAV.

Chapter 6 provides a discussion of airport PPP around the globe.

"PPP is only for big airports"

In many countries, governments have managed to successfully attract private sector participation in airport infrastructure and services. Under a PPP arrangement, the private sector can invest in and operate an airport, improving its safety, security, and the service levels provided to airlines and passengers. Several airports have secured substantial private investment, sometimes without government contribution. In 2010 for example, Pulkovo Airport in St Petersburg, Russia secured more than €1.2 billion of private investment. But demand for private investment is not only for big airports. There is increasing demand from regional airports, not only in large countries such as Brazil, Russia or China,

---

3 See IATA's guidelines on airport ownership and regulation (2018).
4 Kaplan Kirsch & Rockwell, *P3 Airport Projects* (2017).

6 Introduction

but also in smaller countries like Madagascar, Chile and Perú. Concepción in Chile shows how domestic and low-traffic airports can still attract private sector participation to develop a modern airport. Perú successfully transferred several small regional airports to the private sector under a co-financing scheme with the government.

"In time of need, sell the family jewels"

When fiscal space is tight, government budgets are stretched, and the economy has seen better days, there is a temptation to "sell" high value state assets in an effort to "release" value. An airport is a prime target, with good revenues, access to foreign exchange, and a golden future. It is tempting for decision makers to want to sell off an airport. This may not be the wrong decision, but this is the wrong reason to make that decision. Careful analysis is needed. In particular, would the government be better serviced by a share in revenues instead of an outright sale (not to mention control and incentive issues)?

Buy low and sell high: the same logic applies to privatisation. The analysis needs to be done in a dispassionate, careful manner, considering whether to sell now when improvements are needed, or share in the profits later.

"The myth of the hospital pass"

In the great game of rugby, a "hospital pass" involves chucking the ball to a teammate seconds before experiencing a near-fatal tackle by the opposing team. (The tackle is likely to result in a hospital visit.) Some see airport PPPs as the "hospital pass" of the transport sector – a way to offload the difficult and expensive challenges of an airport to the private sector. While PPPs are a good way to get more help resolving such issues, it is worth remembering that the government never steps out of the airport, it merely brings in a partner (hence the name "public-private partnership"). Or, PPP might stand for "preparation, preparation, preparation", requiring careful thought and analysis before commencing the bid process. The government needs to know exactly what it wants, where the risks lie, and how those risks will be allocated before starting a dialogue with private investors.

"The airport authority should enter into a whole series of arrangements, including the PPP, to maintain control and maximise its benefits"

In parallel with designing and implementing a PPP, airport authorities are often tempted to enter into other commercial arrangements for different airport services, for example fuel farms, parking, duty-free, etc., in an effort to maximise its control and revenues earned from the airport. In other cases, the PPP takes time to arrange and the airport authority feels the need to pursue such other commercial arrangements to avoid the appearance of inactivity. But this is not in the airport authority's interests. A PPP operator can deliver comprehensive airport services more efficiently than can numerous, individual service providers. By splitting out services, the airport authority will make less from those services and will achieve less efficiency. The eventual PPP operator will have challenges managing these separate arrangements, and may have to buy them out, which will reduce the revenues available for the airport authority and may undermine the continuity of the entire project. Selling off parts of the airport into different contracts reduces the overall value of the project, decreasing the appeal for some investors/operators. It also increases the perceived risk by having to deal with unknown/unwanted tenants/concessionaires.

"Build it and they will come"

It is commonly believed that after the airport terminal expansion is completed, passenger traffic will increase. But this belief is not necessarily related to capacity concerns. It is a response to wishful thinking: that because there has been an investment, a return may follow. Traffic will increase only if an investment solves an operational restriction on the airside (runways, taxiways, and apron). Stylish new terminal buildings alone will not increase traffic, because passengers are not motivated by an airport to travel, but rather by business, tourism, or a visit to friends and relatives. Have you ever travelled to a city just because the airport was pretty? Or have you ever stopped travelling to a city just because the airport was ugly? The traffic is the response to the market needs, and it exists apart from the airport infrastructure. Investments in airport terminals are driven primarily by the need to provide a good level of service to users (passengers and airlines), and at the same time they serve as a source of national pride.

"Leave well enough alone"

Private involvement is a huge undertaking. It is expensive to prepare, and requires bravery (to address entrenched interests and those less keen to use transparent, competitive procurement). Therefore, some would prefer to avoid private involvement and continue to muddle along. But it is a myth that these difficult decisions can be avoided. When ignored, they grow worse, and more costly. Whether PPP or public reform, these difficult issues need to be addressed.

"It's all about airports"

Public sector airport authorities are often specifically focused on airport functions and their management. This may limit attention to the commercial returns available for airports and associated businesses. Yet PPPs leverage heavily off these commercial revenues. Developing the commercial side of the airport is important to improve the quality of service for the passengers, and to mobilise finance for infrastructure. Decision makers need to understand this dynamic, the detail of how those revenues will be made, and when they should be shared with the government.

"It has nothing to do with the airport" a.k.a. "It's just a shopping mall with airplanes"

The potential for non-aeronautical revenues can transform a marginally profitable airport into a gold mine, but beware the tendency to focus on retail, hotels, conference centres, car parks, or property development. The government needs, first and foremost, a well-run airport. The investor needs to be looking at operating the airport first and making this extra money later. A focus on non-aeronautical operations – in particular during the bidding criteria – can result in the selection of less proficient airport operators, or bids that have not planned well for high-quality airport services. The majority of airports in the world depend primarily on aeronautical revenues.

"It will be a hub"

Policy makers and airport managers often claim that they will attract more traffic to their airport by making it a hub (the obsession to be a hub, or the "hubsession"!). But an airport does not become a hub just by being blessed with a privileged geographical location, or by investing heavily in infrastructure. A hub is not the creation of a policy maker, a regulator or an airport manager. To be a hub, an airport needs to be chosen by an airline that wants to base its operations there. For that to happen, an airport needs an important concentration of origin and destination (O&D) traffic of higher-yield passengers to subsidise the lower-yield connecting traffic. In other words, passengers have the option to take direct flights, and choose routes connecting through hubs due only to lower fares. Passengers are generally willing to pay a premium for the convenience of direct flights. Airlines cannot operate profitably by transporting the majority of their passengers connecting between points other than its base. And the large network of routes generated by the demand of the O&D traffic is what makes it an ideal connection centre for passengers coming from other airports. Without a great deal of the traffic generating from or ending at the airport, and without an airline arriving to exploit that traffic, the airport will never be a hub.[5]

"It will be a cargo hub"

Another common belief is that any under-utilised or abandoned airport can be converted into a cargo hub. Apart from the largest freight integrators,[6] the great majority of world air cargo is shipped in the belly holds of passenger aircraft. It is actually the passenger network system that allows cargo owners and shippers to distribute goods to a variety of destinations. Large cargo operators prefer to be located at passenger airports where they can combine their freight services with passenger belly hold capacity, for thinner routes. The economies of scale required to make a cargo-only airport feasible are present uniquely at a handful

---

5 Unless the operation is government subsidized.
6 FedEx, UPS and DHL.

of airports worldwide, basically at the integrator's hubs. While some perishable goods are often air shipped in large volumes, generating substantial full freighter activity, this is not enough to support the operation of an entire facility. Unless there are substantial levels of imports or exports originating from or destined for a particular airport, or a significant value added at the facility (such as logistics services or some industrial input), the presence of available infrastructure is not enough to develop a cargo airport.

"It will be a low cost carrier airport"

Another elusive, golden-egg-laying goose is the low-cost carrier (LCC) airport. The LCC formula is based mainly on being able to offer low fares by achieving low operating costs. Low cost operations are attained, among others, by low spending at airports. Most LCCs prefer low cost facilities, spend minimum time on the ground, and consume fewer services. They minimise the use of check-in counters, do not contract lounges, avoid using boarding bridges (saving money at the same time as allowing them faster turnarounds), use fewer ramp handling services, load less fuel (because of shorter segments) and purchase very little in-flight catering. Their passengers do not spend much money at the shops, and there is limited dwell time since they don't connect. Ultimately, LCCs aim at point-to-point markets – passengers travelling between city pairs – on a high load factor basis, and as stable as possible throughout the year. Unless the airport can offer large volumes of traffic, the derived revenues from hosting a few LCC flights may be of marginal importance.

"Airports compete with each other"

Governments tend to split up airport groups in different PPPs to "generate competition" between the airports. The truth is that when talking about origin and destination traffic, each airport has its own catchment area, its own market – particularly when other alternative airports are only reachable by traveling long distances by land or sea. With a handful of exceptions around the world where cities are served by different airports that truly compete between them for their passengers and airlines (e.g. Moscow or Tokyo), in most cases passengers do not have a choice. Nor do airlines. Having various airports under one single operator does not hamper competition, because it wasn't there in the first place (see for example BAA and competition amongst the airports around London – or lack thereof). Depending on the individual cases, there might be other considerations to avoid bundling too many airports together, but they have to do with preventing regulatory capture and the balance of power between the operator and the government.

"We will attract MROs (maintenance, repair, and overhaul facilities)"

Among the diverse fantasies many policy makers have is that the development of an airport will be financed by MROs. This is grounded in the belief that dormant runways or airports with very low activity can be used as maintenance facilities to repair airliners. Airlines use MRO facilities to perform maintenance and repair for aircraft. Roughly 40% of the MRO business has to take place at airports, including line maintenance, airframes and modifications, while the remaining activities (engines and components) do not necessarily require the airplane to be flown to the workshop. Airlines prefer to repair their aircraft at airports where they normally fly, so they don't have to ferry the empty aircraft to be fixed. A new trend is to spread major repairs over time, carrying out as many jobs as possible between flights, to avoid having to ground the aircraft for maintenance. Typically, the cost of specially flying an empty narrow body to a workshop could not be offset by any discount that an MRO could offer to its client. Therefore, narrow body operators will seek MROs along their regular operating networks. For wide bodies, only their most comprehensive maintenance checks could justify flying an aircraft out of its normal routes, but only if important discounts compensate for the additional fuel cost, which will then erode the MRO's profitability. That is why MROs are normally found at busy airports with a large established clientele, assuring high margin jobs. The presence of locally based carriers and an important activity of visiting aircraft will not only constitute a captive market for base maintenance (e.g. airframes) but also for line repairs, including unexpected events. In addition, MROs are based where highly skilled workers and competitive labour costs can be easily found, alongside laws favourable to import duties and custom-bonded inventories. Available space and good infrastructure, while useful for an MRO, is not enough to attract MRO operators. They need natural traffic activity, a concentration of home-based aircraft and an important volume of visiting carriers, and the accessibility to skilled and affordable human resources.[7]

"You have to have a minimum level of traffic to use PPP"

There is no minimum traffic for an airport to be considered a PPP; the context of the country and airport will determine what is possible. Most airports under individual concessions have traffic above 500,000 passengers per year but it ultimately depends on the level of investment necessary, traffic, commercial potential and flexibility given to the private sector to run the airport efficiently. Chile has successfully managed to concession airport terminal buildings with just over 600,000 passengers per year. Low traffic airports, that cannot provide the revenues needed alone, can be bundled, using profitability and economies of scale from several airports to achieve the criteria. Another approach is to use management contracts, which can improve significantly the management of any size of airport with limited private investment, and therefore less need for revenues. As another option, the government may prefer to seek private investment, but to also provide support sufficient to overcome such revenue shortfalls.

---

7 With data provided by Eyal Katz from the IAI Aviation Group.

"Flying is for rich people, so public funds should not be invested in airports"

Airports are a critical driver of economic growth, enabling mobility and trade. Aviation's global economic impact – direct, indirect, induced and tourism catalytic – is over USD 2.7 trillion. Aviation supports 3.5% of global GDP. Airports help create jobs: around 6 million jobs globally are directly created by airports, with a total of 62.7 million jobs supported by aviation worldwide.[8] The links between infrastructure and poverty alleviation are well established.[9] Airports improve connectivity. Air travel not only connects people, but it also connects economies; and connectivity drives exports, growth in tourism foreign investment and global economic development. Airports also attract skilled labour and foster economic transformation in an economy. Airports can act as "growth poles" for local economies, becoming destinations of business life, shopping and recreation, and global transport and business centres.

"Building an airport will develop the region"

The construction of an airport will not necessarily develop a region, unless there is a latent and unsatisfied demand for it. A proper demand analysis has to be carried out to determine if the airport will attract new traffic and with it the development of the region. The pressure to lower airfares across the globe motivates airlines to add capacity where economies of scale compensate lower yields with larger volumes. Most greenfield airports replaced old airports, where traffic demand was a given. Completely new airports have been developed, like Ciudad Real in Spain, in the hope of demand that never materialised. In Canada, a second airport for Montreal, Mirabel, became a white elephant after the traffic refused to move out from the more convenient Dorval airport. In Bali, in Indonesia, a new airport is being planned in the north with the justification of economic development, yet it is more likely that current and future demand will continue in the south of the island, and the integration of the north is only a function of an efficient road link.

---

8 Aviation: Benefits Beyond Borders, "Powering Global Economic Growth, Employment, Trade Links, Tourism and Support for Sustainable Development through Air Transport" (2016).
9 See for instance Stephane Straub's working paper on "Infrastructure and Growth in Developing Countries" (2008); César Calderón and Luis Servén's essays on "Infrastructure and Economic Development in Sub-Saharan Africa" (2010a) and "Infrastructure in Latin America" (2010b) as well as their 2011 working paper with Enrique Moral-Benito, "Is Infrastructure Capital Productive? A Dynamic Heterogeneous Approach"; Pierre-Richard Agénor and Blanca Moreno-Dodson's 2006 working paper on "Public Infrastructure and Growth: New Channels and Policy Implications"; Antonio Estache and Grégoire Garsous's 2012 paper on "The Impact of Infrastructure on Growth in Developing Countries"; and Jordan Schwartz, Luis A. Andres and Georgeta Dragoiu's 2009 working paper on "Crisis in Latin America Infrastructure Investment, Employment and the Expectations of Stimulus".

1.2 Key issues

When implementing PPP in airports, government should consider a number of key issues and risks that will need to be managed as part of the PPP, and as part of the government framework responding to PPP. This section outlines a few of these issues, which are discussed in greater detail later in this book.

Land acquisition. The need for access to large amounts of land and space to build transportation facilities makes them expensive, long-term and politically sensitive undertakings. Public reaction to new transport facilities can be challenging, airport neighbours are usually unhappy to see the airport expanding. Generally, land acquisition is at the risk of the contracting authority. It is best for all land to be acquired before the bid process. It may be tempting to try to accelerate the process and commence bidding assuming that the land will be acquired in a timely manner; but this has been the downfall of many a project, and has created massive liabilities for governments across the globe. It took over 15 years for the Government of Perú to hand over the land for a much needed second runway at Lima Jorge Chávez.

Capital subsidies. Transportation projects often involve quite significant capital costs, which may exceed the appetite of the private sector finance market, or the revenue potential of the project while keeping fees and charges affordable, and therefore government support may be essential to the financial viability of the project, whether through capital grants, availability payments or otherwise. While airports may earn revenues sufficient to cover capital costs as well as operating costs, in some cases, in particular for new airports, small facilities or in low-demand scenarios, public support may be needed to contribute to capital costs, in the form of land or also capital subsidies. The €2.1b Athens Venizelos greenfield airport was built under a PPP scheme where the Greek State's equity contribution was 55%.

Commercial issues. Obtaining private investment for an airport raises issues associated with whole business management more than in other transport projects (sea ports aside). Where the contracting authority for a road or bridge will look first for an experienced builder of major civil works, the key investor for an airport is often an experienced operator. Any major PPP airport should require an experienced airport operator as a strategic investor in the consortium.

Revenues. Airport projects involve a multiplicity of commercial arrangements including aeronautical services, terminal facilities and the provision of supporting services such as fuelling, ground handling, cargo warehouses, in-flight catering, parking, hotels, commercial businesses and a variety of other support services which must be supplied by the operator or through a series of different service providers. The contracting authority will need to allow the project company sufficient flexibility in order to improve revenue flow and investment in infrastructure on the site, and the airport experience of passengers. Airport charges can be imposed on passengers and airlines for the use of airport services or other such fees and charges on airside facilities. For large airports, a relevant share of airport revenues comes from non-aeronautical sources or commercial activities of different sources.

Regulation. Careful examination and regulation of private sector involvement in an airport is necessary to avoid substandard design and development as well as appropriate safety and security levels. Some governments impose an architectural design, at least for key features. For example, Pulkovo airport (St Petersburg, Russia) wanted a dramatic roof structure reminiscent of the famous bridges over the Neva river. Not wanting to leave this design to the concessionaire, the City commissioned its design in advance, by a famous architectural firm, and required bidders to include it in their proposals. Any construction or improvements will need to comply with regulatory restrictions and may need regulatory approval. Fees and charges levels in transport are generally regulated to ensure a balance between affordability and cash generation for capital investment. These regulatory matters will influence the timetable for project implementation and should be anticipated at the beginning of the process, in particular where user charges are to be introduced or significantly increased through the PPP project.

PPP process. The process for implementing a PPP project follows a series of stages, from project selection, early assessment, preparation, procurement and implementation. These stages are explored in more detail in Chapters 8–11, and are mapped out in Figure 1.1.

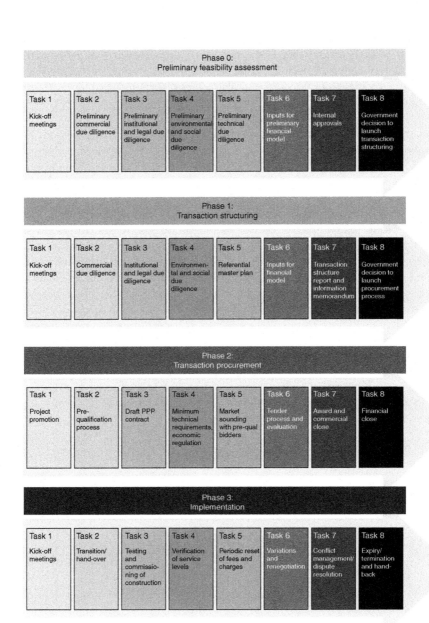

Figure 1.1 The PPP process

2 World trends in air transport

> Life is like riding a bicycle. To keep your balance, you must keep moving.
> Albert Einstein (1879–1955), theoretical physicist and Nobel Prize winner,
> in a letter to his son Eduard (5 February 1930), in Isaacson (2007), p. 367

Air traffic growth has increased over the last 20 years by 5.4% yearly on average. Over this time span the global annual traffic volume has almost tripled (from 1.46 billion passengers in 1998 to 3.98 billion in 2017).[1] World population growth, new technologies and huge emerging markets (like China, India and Brazil) have contributed to this growth. It is expected that in the course of the next 20 years,

---

1 This information is drawn from the World Bank Indicators data as at June 2019 on "Air Transport, Passengers Carried", which is based on information from ICAO, Civil Aviation Statistics of the World and ICAO staff estimates. June 2019.

the annual traffic growth rate for advanced economies will be 2.8% while that of emerging economies will be 5.3%. By 2040, emerging and developing economies will represent 60% of the global air passenger traffic.[2]

Two key factors are reshaping the airline industry at airports around the world: the further expansion of the Low Cost Carrier (LCC) model in the developing world, and the progressive consolidation of legacy carriers.[3]

2.1 The LCC model

The LCC model was born with Southwest Airlines in 1966, out of Dallas, Texas when one lawyer and one pilot (Herbert Kelleher and Rollin King) decided to link three cities, Dallas, Houston and San Antonio, with commuter services, frequent no-frills flights at low fares. The low-cost structure offers very low airfares for short and medium-short haul routes, attracting passengers away from the legacy carriers and competing with surface transportation modes. This success story was later copied in Europe by Ryanair and easyJet, and from there the model expanded worldwide.

The LCC model low cost structure relies on:[4]

- The operation of short sectors (up to about five hours) allows crews to return to their bases, saving accommodation costs
- Simple operations maximise efficiency and allows easier adjustments to enter and exit markets
- Simplified pricing helps to measure results
- Point to point operations, catering to passengers who need to link city pairs. This type of demand is easier to identify and simpler to price. And since there are no connections involved, no derived consequences from delays or cancellations with respect to stranded passengers or misconnection of baggage
- One aircraft type, allowing for commonalities of crews and important savings from economies of scale on aircraft and spares purchasing, training as well as route planning
- Specially crafted labour contracts that are more efficiency driven and allow for multitasking, achieving cost reductions of up to 20% compared to those of legacy carriers
- One-class configuration, with a simpler fare structure and more seats per aircraft
- Greater seat density on each flight, resulting in a lower operating cost per seat
- Quick turn-around at airports, through double door operations, little or no catering and less baggage, facilitating high daily utilisation of the aircraft.

---

2 See the ACI media release: "ACI's World Airport Traffic Forecast Reveals Emerging and Developing Economies will Drive Global Growth", (Airports Council International, November 2018).
3 Also referred as *network carriers* or *full service carriers*.
4 This list is based on Rigas Doganis's *The Airline Business in the Twenty-First Century* (2001).

In addition, LCCs tend to demand fewer services at airports. With a higher proportion of on-line check-in and less checked luggage, they typically require lesser facilities at the check-in areas. At some airports, like Lyon Saint Exupéry, Malpensa Milano, Tel Aviv Ben Gurion or Tokyo Narita, LCCs use separate terminals with less frills and catered retailing for their passengers, as well as lower passenger facility charges.

The affordability of these services has created significant demand, mostly in developed economies. The effect is slowly expanding into the developing world but mostly in domestic services. Regulations affecting international travel tend to result in high costs beyond the control of the airlines, offsetting the advantages of the LCC model. Such is still the case within many markets within Latin America, the Caribbean, Africa and Central Asia.

The LCC model has implications with respect to airport development. LCCs operate out of secondary airports when available (like Ryanair or Wizz Air) or using a different operating model out of main airports (easyJet). Smaller and even inoperative airports have regained traffic after being chosen by LCCs to develop new markets. Airports that were considered a financial burden become profitable and even able to support expansion. Kyiv Zhuliany (Igor Sikorsky) International Airport had around 29,000 passengers in 2010. In early 2011, it was chosen by Wizz Air as an operating base, bringing traffic up to about 1.5m passengers within two years. With further expansion of Wizz Air as well as the addition of about a dozen LCCs, the airport recorded almost 3m passengers in 2018. The former military base El Palomar, in Buenos Aires, went from zero to almost 700,000 passengers in one single year after the arrival of two LCCs, Flybondi and JetSmart.

However, the dream of converting a low operational facility into an LCC base should be considered carefully. Policy makers often imagine the transformation of empty facilities into busy LCC bases, without analysing other dynamics such as market demand, proximity to the catchment area, transportation access and need for investment. The LCC model is based, among other things, on low yields per passenger, so volumes of traffic are essential. LCC operators open services where there is an expectation of high demand that could fill their aircraft achieving high load factors. Airports far from their intended catchment areas may not be suitable for LCCs, in particular when the intention is to move traffic away from the main airports.

2.2 Airport adaptation to LCCs

In the last twenty years, LCCs have transformed the concept of flying not only for passengers but also for airports. Airports had to adapt themselves to the needs of the LCC model by tailoring services and dedicating facilities that are more suitable to LCC operations.

Some LCCs followed a strategy of operating at secondary airports where they could reach better commercial deals of lower fees and charges. Azul Linhas Aéreas from Brazil expanded dramatically out of Campinas Airport (Viracopos), near São Paulo where it could have never found the level of airport capacity required.

While initially LCCs had targeted secondary airports, the business model has matured and shifted its focus, resulting in a scenario where the major share of today's development is taking place in primary airports. For instance, easyJet operates at major airports such as Milano Malpensa (T2) or Lyon-Saint Exupéry (T3), and enjoys dedicated terminals with nose-in parking positions that allow passengers to walk to the aircraft and embark (or disembark) using stairs at the front and rear doors of the aircraft, achieving short turnarounds.

Other airports in Europe that offer dedicated facilities to LCCs are Marseilles, Bordeaux, Sofia (T1), Amsterdam Schiphol Pier H and Copenhagen's "CPH Go". CPH Go is in fact a pier extension of the airport terminal which is only available for specific airlines that meet specific requirements. The terms include maximum turnaround time of 30 minutes and that at least 90% of its passengers must check-in online, via mobile phone or at the self-service kiosks.

In Asia, Singapore Changi introduced a dedicated terminal to LCCs with lower operating costs through fewer facilities, no boarding bridges and commercial outlets with rents up to 50% lower than those at Terminal 1 and 2. Passenger facility charges are also reduced at this dedicated terminal. In 2014 Kuala Lumpur International Airport inaugurated a new terminal (KLIA2) for the exclusive use of Air Asia, where passengers enjoy a reduction of 50% in their passenger service charge.

Tel Aviv Ben Gurion Airport rebranded the old Terminal 1 for LCCs offering a differential service charge to passengers. The terminal is used as a departing terminal only, from where passengers are bussed to the aircraft. Upon arrival, LCC passengers are processed at the main airport terminal (T3).

Dedicated LCC terminals have been a growing trend in the last few years, with further adaptation in progress at many major airports. However, the future integration of short/medium haul operations by the LCCs with the long-haul routes of the legacy carriers will demand further adaptation.

2.3 Legacy carriers

With the proliferation of the LCC model, legacy carriers (the traditional full-service airlines) have had to adapt to a new context of growing competition. In response, legacy carriers have reduced costs by adopting one or all of the following measures:

- Offering low cost features and services
- Forming their own LCC or buying a low cost subsidiary airline
- Establishing partnerships with existing LCCs for specific markets and routes

According to the Centre for Aviation (CAPA), despite these approaches, there is still a meaningful unit cost difference between LCCs and legacy carriers.[5]

---

5 See CAPA's "The Airline Cost Equation: Strategies for competing with LCCs" (2018).

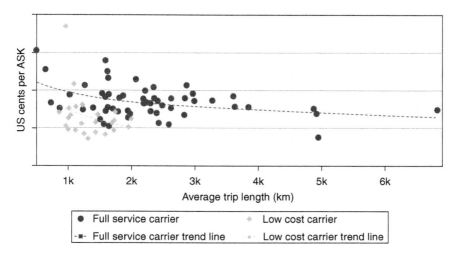

Figure 2.1 Cost per available seat kilometer (CASK, US cents) versus average trip length, world airlines 2016
Source: CAPA (Centre for Aviation), based on airline company accounts (see CAPA 2018)

CAPA argues that legacy carriers usually have higher unit revenues (Revenues per Available Seat Kilometre, or RASK) than LCCs, but there is no evidence that they manage to achieve higher profits as a consequence of this. Quite the opposite – LCCs usually achieve higher margins than legacy carriers through more efficient cost structures.

IATA's projected profitability for 2019 has been downgraded from USD 35.5 billion[6] to 28 billion in light of rising labour and fuel expenses and the cost of infrastructure.[7] Passing the costs through to passengers will be hampered by competition between carriers. In addition, anti-globalisation trends and the US-China trade war is affecting the global air cargo business.

IATA summarises the drivers of global airline profitability as:

- Fuel unit costs will increase as a result of increased fuel prices (from USD 54.9 / barrel Brent to USD 71.6 / barrel Brent in 2019). As a result, fuel costs will increase their share of total airline costs from 23.5% in 2018 to 25% in 2019. Non-fuel unit costs are also expected to increase by 1% in comparison to its 2018 value, due to higher labour and infrastructure costs
- Total airline expenses are projected to increase by 7.4% in comparison to 2018 values
- Global airline revenues are also projected to increase in 2019, but at a slower pace than overall expenses: a 6.5% growth compared to its 2018 value.

6 "IATA Forecasts $35.5bn Net Profit for Airlines in 2019" (IATA, 2018a).
7 "Slowing Demand and Rising Costs Squeeze Airline Profits" (IATA, 2019a).

20 *World trends in air transport*

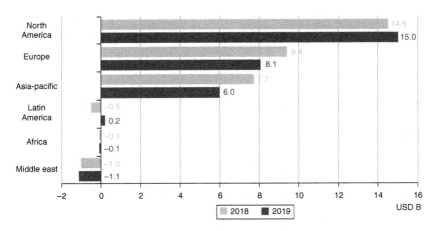

Figure 2.2 IATA revised airline profitability by region, 2019 vs 2018
Source: Based on information from IATA (2019b)

As a result of this evolution, global airline profitability is projected to decrease by 6.7% in 2019. IATA also estimates that North America and Latin America will be the only regions in the world with increased profitability in 2019, with all other regions registering a decrease in net profit.

It is expected that global air cargo volumes will slightly decrease in 2019, as a result of higher tariffs on trade. However, passenger demand is expected to grow by 4.5%, from 4.4 billion global passengers in 2018 to 4.6 billion passengers in 2019.[8]

While some airport charges are transferred directly to the passengers through the air ticket (e.g. passenger facility charges, passenger security charges), charges to airlines are not necessarily passed through to the airfare. In fact, there is ample evidence that the charges that airports levy on airlines are not taken into consideration by the carriers' revenue managers when setting the prices on tickets. Airlines set their prices based on load factors, expected demand and competitors' behaviour and do not consider airport charges as an underlying cost of providing the service.[9]

To confront the progressive reduction of margins, airlines have been joining forces, through alliances, joint ventures and even mergers and acquisitions. This consolidation of airlines has impacted airports, altering competitive conditions and creating mega hubs at the same time as making airlines abandon airports altogether.

8 See again IATA (2019a).
9 See the study by the ICF and released by ACI Europe in 2018 on "Insights into the Logics of Airfares".

The spectrum of airlines has changed dramatically within the last twenty years. In the US alone, from ten main airlines at the turn of the century, four are left today.[10] Something similar has happened in Europe. Three major groups dominate the European legacy carriers: the Lufthansa Group (Lufthansa, Swiss and Austrian; in addition to Brussels Airlines that is fully owned by Deutsche Lufthansa AG); IAG (British Airways, Iberia, Air Lingus, Vueling and Level); and Air France-KLM (including Transavia and HOP, rebranded as Air France HOP). However, in Europe, the LCCs like easyJet, Ryanair and Wizz Air, among other smaller ones, have atomised the market further than in the US. In Latin America, three groups dominate the continent: LATAM (formed by the merger of TAM and the various LAN subsidiaries in Chile, Argentina, Peru and Ecuador); Avianca (by the merger of Avianca Colombia, Grupo TACA that was formed from several Central American airlines and TACA Peru, OceanAir from Brazil[11] and Aerogal from Ecuador); and Copa Airlines (Copa Panamá, Copa Colombia which was previously AeroRepública, and Wingo). Other independent carriers like Aeroméxico, Aerolíneas Argentinas, Gol and Azul, while having extensive domestic networks in their own countries, perform a secondary role in the international connectivity of the subcontinent.

In Asia there seem to be too many airlines. Analysts believe that some consolidation may take place soon. However, large volumes of domestic traffic in countries like China. India, Indonesia, Pakistan, Japan, the Philippines and Vietnam are able to support more airlines with less pressure to consolidate.

The African continent has four main carriers, although they are not the result of consolidation. Ethiopian, South African Airways and Kenya Airways compete intensively in Eastern Africa. At the western part of the continent, Royal Air Maroc is the main carrier, although some spin-off of Ethiopian is expected to develop in this region. The region has witnessed numerous national airlines rise and collapse, after continuous inefficient and non-transparent management.

The effect in the US has been significant; for example, when Delta and Northwest merged, the new airline left Memphis International, previously one of Northwest's hubs. United Airlines left Cleveland as a hub in 2014, after its 2010 merger with Continental. The merger between American Airlines and US Air had a lesser effect on their respective hubs, with only minor shrinkage of operations at Phoenix Sky Harbor.

The effect in Europe has been less dramatic. The airlines from the Lufthansa Group operate independently, each one maintaining its own hub. Within IAG, the merger between British Airways and Iberia may have some consequences on long-haul development at Madrid, Heathrow or Gatwick. The merger between Air France and KLM was followed by a cross shareholding of 8% between Aéroports de Paris and Amsterdam Schiphol Airport, supported by an industrial cooperation agreement.

---

10 American Airlines merged with TWA, which later merged with US Air, which had absorbed America West. Delta merged with Northwest. United merged with Continental and Southwest with Air Tran.
11 OceanAir ceased operations in June 2019.

However, in Latin America consolidation has impacted several airports significantly. The union of Avianca and Grupo Taca that took effect in 2010 significantly benefitted Bogotá El Dorado Airport, which grew steadily at above 6% per year (compounded). At the same time, San José Juan Santamaría in Costa Rica did not see its overall traffic grow after TACA abandoned one of its two hubs, in spite of the increase in traffic by foreign carriers. The same merger affected San Salvador, TACA's main hub, which grew at a significantly lower rate of 4% since 2010.

The consolidation has strengthened the negotiating power of carriers at airports with a lower level of presence. It is common to see airports with dedicated terminals for each of the three large alliances (Star Alliance, Oneworld and Sky Team), such as in Tokyo Narita, London Heathrow and Singapore Changi, to name a few.

Through the progressive code sharing and joint venture agreements airlines could abandon certain airports altogether and redirect traffic through their partners' network.

Any changes in airline strategies should be taken into consideration when developing an airport. Misunderstanding the evolution of the airline industry, many airports embarked on major unnecessary investments.

2.4 The future of the charter business

The charter airline market has historically been linked to leisure tourism travel and tour operators. Originally, the market for charter airline services started as a "pure charter" model, in which flights could only be sold by private tour operators. These operators would sell entire vacation travel packages including the charter flight, by grouping several tourists into a single charter service.

With the appearance and consolidation of large charter airlines, most of them gradually becoming associated or owned by large tourism travel companies (such as the TUI Group and the Thomas Cook Group), the charter aviation business model started to shift towards "scheduled charters", in which tour operators can make reservations for charter flights, and passengers can also book online to pre-scheduled charter services.

LCCs have absorbed part of the charter segment market share in recent years as more people plan their holidays on their own, relying less on tourism agencies. In response, many former charter operators have started selling individual seats as part of their transition to operating low-cost and scheduled services.

During 2018, several important charter airline companies have declared bankruptcy and completely ceased operations: Germania (Germany), Small Planet Airlines (Lithuania, along with its Polish and German subsidiaries), ADI Aerodynamics (United States), OneJet (United States) and Air Link (Australia), and Thomas Cook in 2019.

At the time of writing, the Condor unit of Thomas Cook continues flying through a financing line provided by the German government and is in the process of restructuring.

The number of charter flights has been declining over the past decade: in 2007, they represented 6% of total flights in Europe, the largest market for charter airline services. After ten years, the number of charter flights had reduced nearly by half.[12] Despite the evident slowdown of charter services, market reports and both TUI Group and Thomas Cook Group's financial statements suggest that there has been a slow recovery of the charter market segment in 2017 and 2018.

2.5 Airlines' vertical integration at airports

In order to offer lower airfares to passengers, airlines are disaggregating "ancillary" services and selling them separately, such as checked bags, assigned seats, extra leg room seating and special means (often referred as *à la carte*). Other ancillary services offered by airlines are commission-based products and frequent flyer related products also offered mainly through direct distribution channels.

According to IdeaWorks,[13] between 2007 and 2017 the top ten world airlines increased revenues from ancillary services at an average growth rate of 30%. Airlines with average lower airfares are the ones who achieve the highest percentage of ancillary revenues over total income, particularly the LCCs. While this trend used to be typically found at LCCs (airlines like easyJet, Eurowings, Norwegian and Ryanair can generate as much as 25% of their operating revenues from ancillary activities), recently legacy carriers like Air France/KLM, British Airways and the Lufthansa Group have been implementing basic economy fares on transatlantic routes with ancillary products.

The same trend is occurring in North America where the big three legacy carriers (American, Delta and United) are progressively implementing lower fares, including surcharges for seat assignments and meals. Within Latin America, baggage fees are now permitted on domestic flights within Brazil (Azul, GOL and LATAM offering them). Assigned seat fees are becoming more common across the industry, also in Asia, Africa and Middle East (a trend currently found with the Gulf carriers).

This trend, particularly commission-based products, has had an impact on airports' commercial revenues. Specific services that were traditionally included in the airfare or were purchased at airports on the moment of arrival (car rental, hotel, duty free shops) are offered to passenger by airlines at the moment of purchasing the ticket, or even during the flight.

Airlines consider that the travel experience is not limited to the flight itself, but also includes the airport experience, particularly in the premium segment of products. VIP lounges are now standard at most airports around the world,

---

12 Source: The EUROCONTROL Statistics and Forecast Service (STATFOR).
13 An airlines ancillary revenues specialised outlet based in the US.

a response by airlines seeking to provide the highest yield passengers with an improved travel experience. Particularly at some airports in the developing world, proper seating, air conditioning and respectable washrooms can only be found in the VIP lounges. Either operated by the airlines themselves or under some kind of arrangement with the airport operator, these lounges have now become a commodity. At many international airports, the three major airline alliances now offer differential facilities depending on the frequent flyer tier or the class of booking.

Some airlines provide their higher premium passengers the entire array of terminal services. For example, Lufthansa offers a First Class Terminal at Frankfurt, located outside the main terminal building, with dedicated security and passport control, even including a small duty free outlet. Passengers are taken to the flight by a limousine service, bypassing the airport terminal building entirely.

3 The business of airports

Always behave like a duck – keep calm and unruffled on the surface but paddle like the devil underneath.
Jacob M. Braude,(1896–1970), American judge, humorist and author,
in *Braude's Source Book for Speakers and Writers* (1968), p. 14

This chapter describes the fundamentals of the business of airports, in particular revenues, costs, demand and government support. A PPP must represent a good business opportunity, and therefore these issues will be of critical importance as the PPP is developed. With fees and charges heavily regulated, the potential upside for a PPP airport operator relies on (i) inducing additional traffic,

(ii) increasing the commercial development of the airport, and (iii) achieving efficiencies in operating expenses and capital investment.

When considering PPP arrangements and scope, a good understanding of the airport's commercial potential will optimise the allocation of airport functions, in particular functions that will motivate the private sector and increase revenue for the airport.

For an investor, the intrinsic value of a PPP relies on the potential to increase revenues and reduce costs. Revenues are the result of the traffic throughput and the income derived from each unit of throughput. But costs may be driven by the physical size of the airport, and to some extent by traffic throughput.

This chapter will discuss airport revenues (3.1), with separate discussions on aircraft-related fees (3.2), passenger-related fees (3.3) and different airport revenues contexts around the world (3.4). It will then address airport costs (3.5), the drivers of revenues and costs (3.6), demand (3.7), other key revenue and cost issues (3.8) and finally government support (3.9).

3.1 Revenues

Airport revenues can be grouped as follows:

- *Aeronautical revenues*, typically including proceeds from landing (and lighting) fees, aircraft parking fees, usage of boarding bridges and boarding fees charged to passengers
- *Non-aeronautical revenues* include a combination of:
  - *Ancillary services*: those revenues collected from activities that are provided to airlines but of a commercial nature. The airport may operate these activities directly or charge service providers for the right to operate such services. These services range from ramp handling services, in-flight catering, and aircraft fuelling. The rental of spaces inside the terminal building, such as the use of check-in counters (or terminal equipment), airline offices or lounges, are also typically included within this category
  - *Commercial services*: those the airport collects from tenants, in the form of rent or royalties, dedicated to sell goods and services to passengers and airport users, including duty-free products, food and beverage, retail sales, banking services, transportation services, among others. Some airport operators opt to run these commercial activities directly, through master concessions.

3.1.1 Aeronautical revenues

Aeronautical revenues are derived from the collection of fees and charges for services provided to airlines (and aircraft operators) and passengers which are essential to an airport: providing the necessary infrastructure for the landing, parking and taking-off of aircraft, and providing the infrastructure and the service of processing passengers. Typically, they include proceeds from landing fees, aircraft parking fees, usage of boarding bridges and the boarding fees levied on passengers (also called "passenger facility charge",– PFC, or "passenger service charge" – PSC). Many airports around the world introduced a "security charge" after the events of September 11. Other sources of revenues include runway lighting fees or specific charges for operating beyond normal hours, aircraft approach fees (if not collected by the air navigation service provider), and specific surcharges for operating during peak times or when using environmentally sensitive equipment (noise or emissions).

Aeronautical revenues are the core of any airport PPP, representing typically more than 65% of the total revenues, and at small facilities could be the only source of revenue. A common trend is that all fees and charges to passengers are levied together with the air fare, in the ticket price. The overall fees and charges can represent a significant portion of the total ticket price, which can potentially affect demand for air travel. One measurement of the success of a PPP transaction is the improvement in the level of service to passengers at no or relatively limited extra cost to passengers or airlines.

Fees and charges for aeronautical revenues are normally regulated, thus safeguarding consumers against any abuse that may result from the monopolistic power of the airport operator, since there is no market power through competition. The regulations preclude the private operator from increasing fees and charges at their own will, while allowing some kind of adjustment considering inflation and the level of investments. Fees and charges are typically set by laws, decrees, or other regulation. Fees and charges are usually adjusted through automatic or ad-hoc mechanisms, and could be linked to the completion of specific investments. The share of aeronautical revenues depends on the regulatory scheme, as described in Chapter 5.

3.1.2 Non-aeronautical revenues

The greatest opportunity for value generation is non-aeronautical revenues, from services that are not aeronautically essential to an airport but are commonly found there.[1] Non-aeronautical activities comprised of ancillary services and commercial activities.

Ancillary services include:

- Attending aircraft, for example, ramp handling, towing, push back, cleaning, loading and unloading of baggage and cargo, and provision of ladders, waste removal, electric power, potable water, etc.

---

1 One could conceive of an airport without commercial services, and even without fuel, ramp handling or inflight catering services. However, an airport would not be one if it does not have a runway, a taxiway, and apron and some kind of structure that serves to process passengers. In fact, most regional aircraft are equipped with deployable stairs to be able to do without most of the usual ramp handling.

28 *The business of airports*

- Passenger services, for example, baggage handling, check-in counter services, lost baggage tracking, check-in services, lost & found, concierge services, and access to the CUTE system[2]
- Supplying fuel to airlines – into-plane fuelling
- Providing in-flight catering to airlines
- Renting spaces to airlines – VIP lounges, crew-briefing rooms, ticketing offices, etc.; and
- Leasing space to companies supplying inflight catering services to airlines.

These services are usually provided through third party operators that pay an access charge in the form of revenue sharing to the airport landlord through a royalty, a rent or a combination of both, generally collected on an area basis, or on a per-passenger or time basis.

Revenues from ancillary activities are often regulated, either by contract or regulatory framework. They may also respond to market pricing. For example, the access charge levied by the airport on service providers is normally regulated to prevent abusive behaviour by the airport operator. Costs for the services to airlines might be regulated at some airports, where competition of providers is not sufficient.[3]

These services are provided through a wide range of models. Some of these services may be provided exclusively by the airport operator (e.g. ramp handling services at Frankfurt), or by a subsidiary with exclusive rights (e.g. ground handling services at Kathmandu Tribhuvan), by one sole vendor with exclusive rights to provide to airlines (e.g. ramp handling at Buenos Aires Ezeiza[4]), or by granting freedom to any provider that complies with safety and security norms (e.g. most services at Athens Eleftherios Venizelos). Granting the right to third party providers to operate is set by access charges, collected by the airport in the form of royalties (calculated as a percentage of gross sales, or a fixed charge per unit of service, or even as a rent per area occupied). In order to be able to collect fees from airlines providing the service for themselves, some airports have implemented a fixed value per type of operation, regardless of the provider (e.g. Lima Jorge Chávez).

Commercial revenues comprise all proceeds from commercial activities provided to passengers and to the general public, including *meeters* and *greeters*, airport and airline employees, visitors and the neighbouring community. Activities include selling food and beverages, general retail, banking, communication, transportation services, duty free shopping for international passengers and car parking, among others. Some airports include, under this category, revenue from real estate development. Typically, these activities are carried out by tenants under

---

2 CUTE refers to Common Use Terminal Equipment, used by airlines to check-in passengers. And there is nothing cute about it.
3 As per EU directive (Council Directive 96/67 /EC), airports with less than two million passengers should allow the provision of ground handling by third parties and airports with less than one million passengers should allow the freedom of self-handling.
4 Its real name is Ministro Pistarini but we like to call it by the name of the town.

payment arrangements that take the form of royalty fees (calculated as a percentage of gross revenues), a fixed rent (based on the area provided) or a combination of these.

Commercial activities at airports were limited to food and beverage until the late 1960s. With the expansion of the duty-free shopping and the change in the commercial dynamics of airports globally, the scope of non-aeronautical activities has expanded significantly.

Together with the efficiencies gained in operations, the increase in commercial revenues is often the strongest incentive to seek private sector participation. Airports provide a variety of opportunities to supplement aeronautical revenues with commercial revenues. Different models include layering in property development and other uses of airport property for revenue generating activities. These commercial assets may need to be disaggregated for regulatory or legal reasons, or where commercial activities need to be depreciated over a different period.

Large airports with important volumes of traffic have managed to generate significant revenues from non-aeronautical revenues), reaching 50% of total airport revenues (e.g. Sydney). However, the potential for increase in revenue from non-aeronautical activities varies significantly among airports, and has a direct relationship to volume throughput and the type of traffic. Governments should assess carefully the feasibility of generating additional value from these activities in addition to value generated from aeronautical activities.

3.2 Aircraft-related fees and charges

The fees and charges applied to airlines for airport services cover a number of different activities, including landing, aircraft parking, boarding bridges and terminal navigation.

3.2.1 *Landing fees*

Aircraft landing fees are collected for the usage of runways and taxiways (at some airports, an additional lighting charge may be levied for night-time operations). Landing fees are charged per operation (at almost all airports in the world, only one fee is charged for a combined landing and subsequent take-off).

In general terms, landing fees are levied based on the aircraft's Maximum Takeoff Weight (MTOW)[5] by multiplying the tonnage of the aircraft by a fixed rate.

The methods for charging differ by regions. While at airports in one region the charges are calculated in a straightforward manner, multiplying the weight by the

---

5 The Maximum Takeoff Weight of an aircraft is a fixed value defined by the aircraft manufacturer in its Certificate of Airworthiness. This value is the maximum mass at which the aircraft is certified for takeoff, and cannot change with variations in temperature, altitude or runway conditions.

30 *The business of airports*

unit rate (JFK), at others weight bands are used to differentiate unit rates per ton (Santiago de Chile Arturo Merino Benítez Airport).

Airports like Frankfurt, Heathrow and Vienna incorporate additional noise-related charges to promote aircraft noise reduction efforts.

Some airports construct the landing charge for aircraft by using both a fixed value per landing (common to all aircraft types), plus a variable component to be multiplied per each aircraft MTOW ton. Such is the case of Paris Charles de Gaulle International Airport, as shown in Table 3.1.

3.2.2 Aircraft parking fees

Aircraft parking fees are usually levied on airlines for the time their aircraft spends on the ground between flights. It may also be related to the aircraft's MTOW (with higher fees levied on larger aircraft), or calculated as a percentage of the total landing fee (e.g. Lima Jorge Chávez Airport), which also makes it related to the weight of the aircraft. In most airports, there is a grace period for free parking that varies across airports, ranging from 90 minutes to 4 hours. For these periods the parking is included in the landing charge and beyond the grace period, charges could go by the hour or jump to a daily rate.

Some airports charge different parking fees according to the place where the aircraft is located, usually with higher fees for aircraft parked at the apron in front of the terminal than for those standing at remote aprons. Airports with a large proportion of long-haul night flights receive aircraft in the morning that depart again only in the evening. The planes are normally removed from the main apron in front of the terminal and towed away to a more distant ramp, at a lower parking rate. At any given day during noon time, it is common to see a large number of wide bodies from the major European carriers parked at the northern apron of Johannesburg O. R. Tambo Airport, after a morning arrival awaiting the evening return. A similar situation can be seen at São Paulo Guarulhos, where a large number of US and European carriers arrive during the morning hours and do not depart until the evening. At both of these airports, parking charges at the remote aprons are lower than at the terminal.

3.2.3 Boarding bridge fees

Boarding bridge fees are levied for the time the aircraft is connected. After an initial period of connection the intervals tend to be shorter, to provide an incentive to airlines to remove the aircraft from the boarding bridge as soon as they have finished using it.

However, some airports consider not only the connection time but also the aircraft's MTOW to calculate boarding bridge fees. Athens Eleftherios Venizelos, Skopje Alexander the Great, Belgrade Nikola Tesla and Prague Vaclav Havel are some examples of this practice.

Table 3.1 Example of landing charges structures by region, 2019

Region	Most used mechanism for the collection of landing fees	Examples
North America	Fixed fee per 1,000 pounds of MTOW	New York JFK: USD 6.27 per 1,000 pounds Los Angeles LAX: USD 10.47 per 1,000 pounds
Latin America	Weight bands	Santiago de Chile SCL: (fee per MTOW ton): Aircraft below 49 tons, USD 2.76 Aircraft with MTOW between 49.1 and 89 tons of MTOW, USD 4.12 Aircraft above 89 tons of MTOW, USD 4.69
Europe	1) Fixed fee + fixed fee per MTOW ton 2) Fee per enplaned passenger + fixed charge per noise category 3) Fixed fee + weight bands	1) Paris CDG: fixed fee of EUR 295 + EUR 4.12 per MTOW ton (>89tons) 2) Frankfurt FRA: EUR 1.36 per enplaned passenger + fixed fee (e.g. for an A320, noise charge is of EUR 402 per turnaround) 3) Rome FCO: fixed charge per movement of EUR 55.72 (EUR 111.44 per turnaround) + variable fee per MTOW ton, using weight bands
Asia	1) Fixed fee + variable fee per MTOW ton 2) Fixed fee + variable fee per MTOW ton, using weight bands	1) Hong Kong: fixed fee of HKD 3.150 + HKD 74 per MTOW ton (in excess of 20 tons) 2) Bangkok: For a B737-800: fixed fee of THB 6,550 + THB 155 per MTOW ton in excess of 50 tons For a B777-200: fixed fee of THB 14,300 + THB 175 per MTOW ton in excess of 100 tons
Africa	Incremental weight bands	Addis Ababa ADD: MTOW up to 5,000 pounds: USD 5.86 MTOW btw 5,000 / 40,000 pounds: USD 5.86 + USD 1.75 per 1,000 pounds MTOW above 40,000 pounds: USD 67.1 + USD 2.64 per 1,000 pounds

Sources: DGAC Chile, Port Authority of New York and New Jersey, Los Angeles World Airports, Aéroports de Paris, Fraport, Roma Fiumicino Airport, Hong Kong Airport Authority, Civil Aviation Authority of Thailand, Ethiopian Airports Enterprise

32 The business of airports

Table 3.2 Example of parking charges structure, 2019

Weight band	Fee per each MTOW ton, per hour
Aircraft below 100 tons of MTOW (first 2 hours free of charge)	EUR 0.75
Aircraft above 100 tons of MTOW (first 4 hours free of charge)	EUR 0.75

Source: Brussels Airport

Table 3.3 Example of boarding bridge charges structure, 2019

For all aircraft types	Up to 45 minutes of connection (in USD)	Charge per additional 15 minutes (in USD)
Per connection	128.24	42.75

Source: Lima Jorge Chávez Airport

3.2.4 Terminal Navigation Charge (TNC)

Terminal Navigation Charges cover services provided by tower and approach controllers to aircraft arriving at a location where a tower service is in operation. These include separation of aircraft in tower airspace under radar and visual surveillance, in approach airspace under radar surveillance, and under procedural rules where there are no surveillance technologies.[6] In addition, these services also include the provision of navigation aids at the airport (such as ILS, VOR, NDB, or DME[7]) and information for surface movement separation in runways and taxiways, as well as traffic information in aprons.

It is customary that these charges are collected by the air navigation service provider, normally used to fund the infrastructure dedicated to this service. However, some airports include this charge as part of their aeronautical income for providing and maintaining the equipment.

3.3 Passenger-related fees and charges

The fees and charges applied to passenger activities include passenger service charges, security fees and specific development fees.

---

6 See here Airservices Australia's 2010 discussion paper on "Terminal Navigation Pricing Review", available online.
7 ILS: Instrument Landing System, comprising series of antennas to assist landing under minimum visual conditions. VOR, NDB and DME are different types of navigational aids assisting in the location and approach of aircraft.

3.3.1 Passenger Service Charge (PSC)

As discussed above, a Passenger Service Charge (or Passenger Facility Charge) is levied on departing passengers, related to the use of the terminal building facilities and infrastructure and to the handling of luggage. It is a fixed fee that is not related to the value of the airfare paid by the passenger, and it is collected by airlines at the moment of purchase of the air ticket (and then forwarded by the airlines to the airports). It is customary to find differential PSCs for international and domestic flights, because of the different scope of services required by passengers.

3.3.2 Security fee

Security fees are also levied on a departing passenger basis. They are collected by airports to obtain funding for the provision of security services, including the screening of passengers and their luggage. Some airports include the provision of security services within the PSC, without charging an additional fee.

3.3.3 Specific development fees

Some airports may add an additional fee for the development of specific infrastructure, usually for a limited period of time and calculated as a fixed value per enplaned passenger. Examples of this type of fees include Athens Eleftherios Venizelos International Airport Development Fee, Kingston Norman Manley International Airport Improvement Fee, and Punta Cana International Airport Infrastructure Fee, to name some.

3.4 Non-aeronautical revenues around the world

From an analysis of over 200 airports, non-aeronautical revenues vary by region, because of the consumption patterns and the composition of different levels of economic development.

Asia-Pacific has the highest average non-aviation revenues per passenger, which is explained by consumption patterns of luxury brands and duty free products. Latin America registers the lowest revenue per passenger, mostly due to the lower commercial revenues recorded in US dollars (Figure 3.1).

When analysing these performance indicators according to airport size, three different groups were identified:

- Airports with less than 5 million annual passengers (classified as "small")
- Airports with more than 5 million but less than 10 million annual passengers (classified as "medium")
- Airports with more than 10 million annual passengers (classified as "large").

As a result, Key Performance Indicators have been obtained for each of these groups.

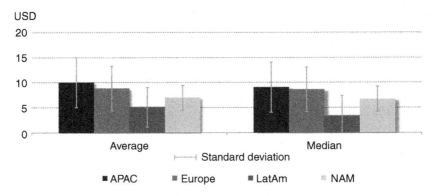

Figure 3.1 Non-aeronautical revenues per passenger by region, 2017
Source: Based on information obtained from financial statements, annual reports and traffic reports published by 108 airport companies and airport groups for 2017, representing over 220 airports
Note: APAC = Asia Pacific; LatAm = Latin America, NAM = North America

Figure 3.2 shows that large airports have higher sample average values for non-aviation revenues per passenger than those of medium and small airports. Commercial facilities tend to be more developed and offer more possibilities for consumers at large airports. The analysis shows that average non-aviation revenues for large airports is USD 8.1 per passenger, a value that is 25% higher than that of medium airports and 90% higher than that of small airports.

Considering sample average values, large airports show significantly higher total revenues per ATM (USD 2.371) than those of medium (USD 1.687) and small (USD 1.065) airports. This is partially explained by the fact that larger airports usually have a higher proportion of operations performed with wide-body aircraft, which implies larger aeronautical fees and charges than ATMs performed with narrow-body aircraft. This is corroborated by the data obtained for aviation revenues per ATM indicator: large airports average USD 1.288 per ATM, while medium airports average USD 1.019 per ATM and small airports USD 764 per ATM.

3.5 Costs

Airport costs include capital expenditures (capex) and operating expenses (opex).

3.5.1 Operating costs

Controlling the operating expenses of the airport is one of the major sources of value generation for most PPP investors. Typically, the operating costs of an airport should range from 40% to 55% of the operating revenues. Airport costs are

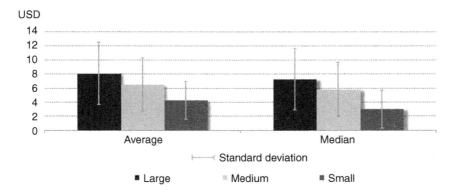

Figure 3.2 Non-aviation revenues per passenger by airport size, 2017
Source: Based on information obtained from financial statements, annual reports and traffic reports published by airport companies and airport groups

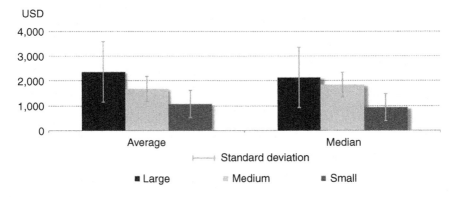

Figure 3.3 Total revenues per ATM by airport size, 2017
Source: Based on information obtained from financial statements, annual reports and traffic reports published by airport companies and airport groups

related to the physical size of the facility and the number of employees and, to a lesser extent, traffic volume.

The largest categories of airport costs are labour, maintenance and cleaning, service contracts[8] and utilities. The size of airport facilities determines the number of employees, and the cost of maintenance. For example, lawn cutting represents an important cost for many medium and small airports, with no relationship to the level of traffic.

---

8 Service contracts could be any outsourcing of labour or maintenance. The information published by airports does not provide sufficient detail to differentiate the nature of the contract.

Although the number of landings at a runway may represent costs for tyre-rubber removal at touch-down zones, a great deal of pavement maintenance costs relate more to temperature variations and precipitation than operations. Terminal cleaning contracts are often negotiated on a per area basis even if there is some relationship between the cost of cleaning and passenger volume. Similarly, security screening costs are usually linked to the number of gates rather than to the number of boarding passengers.

Utilities, such as power supply, are more likely to be a direct function of the hours of operation of the airport, the season of the year as well as the level of traffic. Climate control, lighting and electromechanical equipment are the largest consumers of electric supply.

The scope of airport activities may also change the cost structure significantly from airport to airport. While some airport companies only operate the terminal building (for example, Liberia Airport in Costa Rica), others may also operate vertically integrated services (Fraport airport operating ramp handling services). Some airports operate their own duty-free stores (Dubai International) while others outsource it to specialised international companies (Zurich airport outsources it to Dufry). Similarly, car parking facilities are independently run at some airports (Heathrow), while others are outsourced (Helsinki).

In all of these cases, the different ownership schemes create significant variations of the associated costs for maintenance, labour and utilities. When either benchmarking across airports or measuring indicators to extract drivers, the peculiarities of each case should be carefully considered.

The distribution of these costs depends on the level of outsourcing of services. Maintenance and cleaning services are among those that airports tend to contract out. An airport that outsources services through service contracts will show a lower proportion of labour costs than would airports with similar levels of traffic that does less outsourcing. For example, security services, particularly passenger and carry-on baggage screening, might be outsourced to private companies (like G4S at Barbados Grantley Adams International Airport, providing passenger screening), or may be the responsibility of the central government (like Transportation Security Administration, at all US airports). At other airports, like Tel Aviv Ben Gurion, the majority of security staff is airport authority personnel.

To project future costs of an airport, we must identify the drivers of future cost escalation, including the relationship between these drivers. For example, the cost of cleaning a terminal building is directly related to the number of passengers that pass through. If the terminal is to be expanded, or the number of passengers will increase, the cost base may grow differently.

The challenge of a PPP operator is to gain efficiencies in each one of the cost items. Typically, the new airport company will review the service contracts and renegotiate them under better terms. The consumption of utilities (such as electricity, fuel and gas, water and disposal of waste) is re-evaluated, looking for alternative sources of power and more efficient methods of waste disposal.

The introduction of new equipment (e.g. conveyor belts, baggage handling system, escalators, lifts, etc.) has implications with respect to maintenance, while a more efficient management will switch from corrective to preventive maintenance.

Payroll costs will depend on the decision of the airport operator to have either internal staff to deliver the airport's operation and maintenance or to outsource the majority of these services to third party organisations. For example, according to London Heathrow Airport's financial statements for 2018, salaries and benefits represented 35% of the airport's total operating expenses, being its most important cost category,[9] in contrast, at Amsterdam Schiphol Airport payroll costs represented 23% of its total operating expenses, being largely surpassed by outsourced service contracts, which represented 65%.[10]

Considering the financial information of a sample of 93 leading airports and airport groups across the globe for 2017, the average composition of operating expenses per category is shown here in Figure 3.4.

Despite the fact that salaries and benefits and outsourced services represent a combined average of almost 70% of airports' total operating expenses, the distribution over cost categories varies significantly from case to case. Figure 3.5 presents the distribution of operating expenses among four main cost categories at 25 airports and airport groups across the globe, for 2017.

There seems to be no clear trend among the 25 airport companies depicted in these two figures; each has a different cost composition based on their chosen management structure.

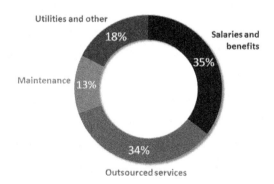

Figure 3.4 Typical composition of an airport operating expenses
Source: Based on financial statements published by 93 airports and airport groups

---

9 Heathrow Airport Holdings Limited, "Annual Report and Financial Statements for the Year Ended 31 December 2018".
10 Royal Schiphol Group, "Annual Report 2018".

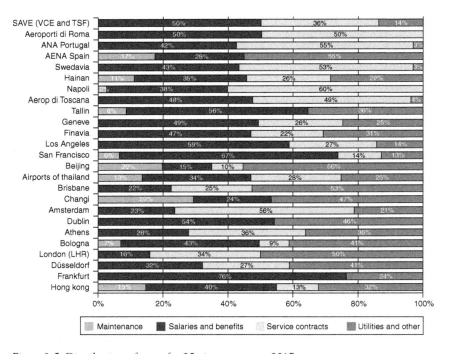

Figure 3.5 Distribution of opex for 25 airport groups, 2017
Source: Based on financial statements published by 25 airport companies, over 80 airports

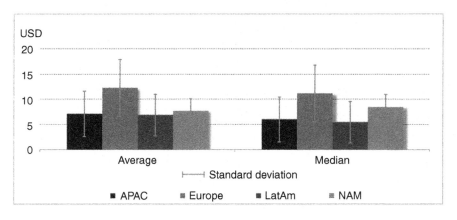

Figure 3.6 Operating expenses per passenger by region, 2017
Source: Based on information obtained from financial statements, annual reports and traffic reports published by airport companies and airport groups
Note: APAC = Asia Pacific; LatAm = Latin America, NAM = North America

The business of airports 39

As seen in Figure 3.6, Europe is the region with the highest operating expenses per passenger, with a sample average value of USD 12.3 per pax. The remaining regions have considerably lower sample average values of between USD 6.9 and USD 7.7 per pax. Higher and more homogeneous income levels in European countries could be a large contributor to these differences. Some airports/airport groups in the Asian region are also located in high income level countries, but others in that same region are located in lower income level countries (such as Malaysia Airports, Airports of Thailand, Beijing, Hainan).

The sample average of operating expenses per pax at North American airports is of 7.7 USD per pax with a standard deviation of 32% of this value (the lowest among all regions), showing the robustness of this KPI for this region.

The same dynamic is observed for operating expenses per ATM (Figure 3.7): the sample average of European airports is almost three times as large as that of Latin America and twice as large as those of Asia Pacific and North America. The low value observed for Latin America is mostly explained by low income levels of the countries in this region.

EBITDA margin sample averages are between 55% and 64% at APAC, Latin America and North America, with the European sample average being 42%. It is remarkable that airports in North America have registered a sample standard deviation of only 10% of the sample average, showing the robustness of this indicator for this region. EBITDA margins show significant dispersions in the remaining regions, particularly in Europe (where the sample standard deviation value represents 43% of the sample's average value) and Latin America (37%).

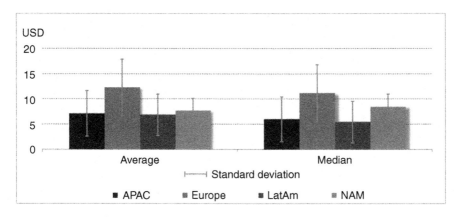

Figure 3.7 Operating expenses per ATM by region, 2017
Source: authors, based on information obtained from financial statements, annual reports and traffic reports published by airport companies and airport groups
Note: APAC = Asia Pacific; LatAm = Latin America, NAM = North America

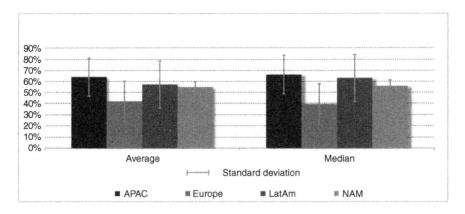

Figure 3.8 EBITDA margin by region, 2017

Source: Based on information obtained from financial statements, annual reports and traffic reports published by airport companies and airport groups
Note: APAC = Asia Pacific; LatAm = Latin America, NAM = North America

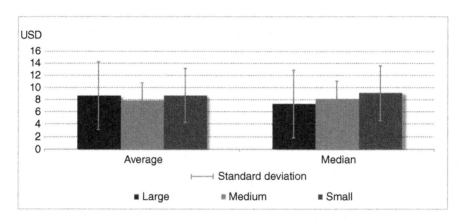

Figure 3.9 Operating expenses per passenger by airport size, 2017

Source: Based on information obtained from financial statements, annual reports and traffic reports published by airport companies and airport groups

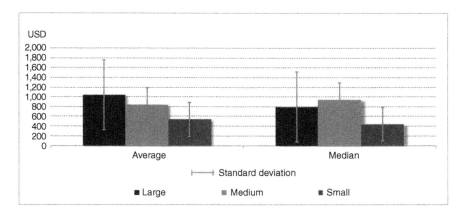

Figure 3.10 Operating expenses per ATM by airport size, 2017
Source: Based on information obtained from financial statements, annual reports and traffic reports published by airport companies and airport groups

If only sample averages were to be considered, operating expenses per passenger would not seem to be correlated to airport size. Large airports have a similar sample average as those of medium and small airports. Values for all airport types are within the USD 7.9 to USD 8.7 per passenger range.

However, as sample average standard deviation values are significant for large and small airports, it is also important to consider sample median values: using this criteria, large airports have a sample median value of USD 7.4 per passenger, which is 20% lower than that of small airports and 10% lower than that of midsized airports. These results suggest the existence of economies of scale at larger airports, measured at a per passenger level.

Small airports have a sample average value of operating expenses per ATM (USD 542) that are 48% lower than large airports and 36% than midsized airports. The fact that small airports have a higher proportion of operations with regional and narrow body aircraft than medium and large airports is a key factor that explains these differences, as operations with smaller aircraft imply less operating expenses per ATM.

Even when the sample average for large airports is higher than that of medium airports, the high value of the average sample standard deviation of large airports makes it important to also consider median values: when considering this parameter, large airports have a smaller sample median per ATM (USD 801) than that of medium airports (USD 944). This suggests the existence of economies of scale at the majority of large airports, with some discrete data points at elevated operating expenses per ATM, elevating the level of sample dispersion.

Large airports have a sample average EBITDA margin of 56%, surpassing that of mid-sized (51%) and small airports (46%). The higher level of profitability at

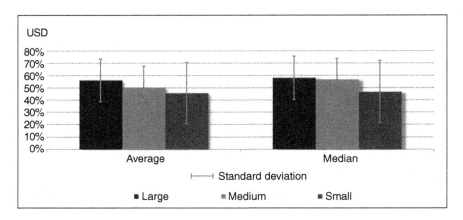

Figure 3.11 EBITDA margin by airport size, 2017
Source: Based on information obtained from financial statements, annual reports and traffic reports published by airport companies and airport groups

large airports seems to be associated with the existence of economies of scale in operating expenses (measured at a per passenger level) and increased consumer spending derived from enhanced commercial areas.

> **Box 3.1 Labour**
>
> PPP can bring new and better conditions for personnel. A PPP investor will generally only bring a few managers, and otherwise use local staff, if they are good. More replacement may be needed where there is a culture of corruption or where staff refuse to change. Those employees who stay are generally happier, better paid and most importantly, properly trained.
>
> However, in some countries, state owned airports are used for political motives to create jobs, placing a financial burden on the airport, limiting investment and often stripping asset value. While reforms can release airport value and improve services, they may threaten the livelihoods and advantages created.
>
> Reform processes need to be consulted with employees, labour unions and the local community to help manage these concerns and the impact of any proposed reform agenda.

3.5.2 Capital expenditure

Capital expenditures (capex) or investments at airports can be divided into three categories: (i) to comply with safety and security standards, (ii) to remediate the consequence of deferred maintenance, and (iii) to expand the facilities to cope with current or future demand.

(i) Investments related to compliance

Airports have to comply with local standards and regulations with respect to safety and security for all facilities at the airport, including the airside (runways, taxiways and aprons) and landside (terminals, terminal access, etc.). According to international regulations, any country that is an ICAO member state should adopt into its own legislation all ICAO standards and recommended practices (SARPs).[11]

Applicable regulations change over time, creating a particular challenge for airport compliance investments. For example, after a series of accidents and incidents, in 1999, ICAO increased the required length of the runway end safety area (RESA)[12] forcing airports around the world to comply with the new standard by extending the runway length to accommodate the new safety area. In order to meet this norm, Kingston Norman Manley International Airport included a 300 metre extension of the runway as a mandatory investment in its PPP concession of the airport in 2018. This requirement involved a major investment since the extension will have to be carried out into the bay.

New investments may result from changes in patterns of operation. For example, the introduction of services by larger aircraft may require the presence of a higher category of protection of rescue and firefighting services, and therefore investment in new vehicles and associated equipment.

An assessment of compliance should result in a prioritised list of investments according to level of urgency. For example, an airport's runway may not comply with specific norms of lighting, or separation from specific obstacles, or the perimeter fencing of the airport may not have an associated road for proper inspection and patrol. The prioritisation of investments will help ensure the greatest level of compliance possible in the shortest time.

(ii) Investments derived from deferred maintenance

When preparing a PPP, investments may be required to compensate for past neglect of important maintenance works. For example, the runway pavement may suffer from structural defects or resurfacing may be needed immediately. While maintenance works are considered regular operating expenditures, major works for deferred maintenance are usually considered as capital expenditures and incorporated in the investment program. For example, when the French consortium Ravinala took over the airports of Antananarivo Ivato and Fascene Nosy Be

---

11 Please refer to Chapter 5: Regulation.
12 Thus converting a "recommended practice" into a "standard".

in Madagascar in December 2016, major repairs on the runways were required as initial mandatory investments because of deferred maintenance.

(iii) Investments related to expansion of facilities

An airport may need investment in expansion, either on the airside and/or the landside to address existing or future expected demand. Based on the projected demand, a master plan will determine the required size of the facilities needed to accommodate anticipated traffic. The master-plan will forecast investment over 30 to 40 years, but should be revised every five years to adjust for actual growth in demand and for technological changes. Capacity constraints might be a limitation for growth in the runway system or the number of available aircraft parking stands at the apron, for example. Failing to expand the terminal building to accommodate demand may reduce the level of service because of inadequate facilities given the increased volume of passengers.

The way future capacity requirements are estimated for air traffic movements and for passenger growth is different. Airside facilities, such as the capacity of the runway and taxiway systems, depends on the maximum requirement of landings and takeoffs at the highest level of demand that is expected, called the *peak-hour*. The required number of aircraft stands (or parking positions) depends on the maximum number of aircraft that are expected to be present at one time at the apron. This number depends not only on the level of throughput of the airport but also on airline schedules, how much time each aircraft will spend on the ground between flights, and those that stay overnight.

Changes in the operational plans of an airline, carriers based at the airport and regulatory issues such as night curfews are among the various factors that may dictate the required number of parking positions needed.

The planning of landside areas, such as the passenger terminal building, must consider the expected volume of passengers anticipated according to the desired level of service. However, facilities are not designed for peak hour, but for a specific volume of throughput according to the *design hour* (or the busy hour). Just as a church is designed for a regular Sunday and not for the crowds of Christmas Eve,[13] a terminal building will not be designed to accommodate the number of passengers expected on the busiest days of the year, like the beginning of the school holidays. Instead, there are different methods to define the design hour of a terminal building. IATA defines the design hour as the second busiest day of the average week of the peak month of the year. The FAA, instead, uses the Peak Profile Hour (PPH), calculated as the peak hour of the average day of the peak month. The Standard Busy Rate (SBR) employed by the UK Civil Aviation Authority, considers the thirtieth busiest

---

13 Or for that matter, a synagogue is not designed for Yom Kippur but for a regular Shabbat, and a mosque is not designed for Ramadan but for a regular Friday.

hour of the year as the design hour for which facilities should be planned. The French (Aéroports de Paris) use the fortieth busiest hour, the Dutch (Schiphol) use the twentieth busiest hour, or the ninety-fifth percentile of the total annual traffic.

Once the design hour is defined, and according to the desired level of service,[14] airport planners, architects and engineers determine the required level of investment required. The required investment in facilities for a new airport must include the required level of finishing and specific design desired for a terminal building. For example, during the planning of the expansion of the international passenger terminal at St Petersburg Pulkovo Airport for its PPP in 2009, the City of St Petersburg specified a particular roof for the new terminal building, designed by the famous British architect Nicholas Grimshaw.

3.6 Revenue and cost drivers

Aeronautical revenues are sensitive to regulatory changes, including mechanisms to set and adjust fees and charges, and the right to collect fees and charges.

Since aeronautical fees and charges are normally set rigidly by regulation, the airport operator will have little influence over them. Typically, airports have no control on these types of charges. In some cases, airports may not even have the right to collect them directly, but through some government agency. In 2009, the Ecuadoran constitutional court found that the financing plan for the 2002 new Quito International airport was partially unconstitutional and that the concessionaire's right to collect passenger charges was not legal.[15]

Non-aeronautical revenues, while related to the passenger throughput, may not have a direct relationship therewith. For example, airports deriving an important share of revenue from selling services to the local community may not necessarily see revenues decrease sharply during traffic shortfalls. Bucharest Otopeni, during the 1990s, had the largest supermarket in the city. For many years, the restaurant at Rio de Janeiro's domestic airport of Santos Dumont was a fashionable place for downtown executives. Airports with a large number of employees also enjoy high commercial revenues from the local community – every day, over 80,000 employees spend their day at Frankfurt Airport. Santiago de Chile airport saw its commercial revenues grow during periods of bad weather, with aircraft on the ground and thousands of passengers stranded and consuming food and beverages.

Another example would be revenues from ramp handling companies. A reduction in the load factor means aircraft flying with fewer passengers, but not necessarily affecting the number of flights.[16]

---

14 Like the IATA LoS, as defined by the ADRM (see Chapter 5 on Regulation).
15 See IATA's guidance booklet on *Airport Ownership and Regulation* (2018).
16 The opposite case may also be true: if the aircraft are flying with greater load factors, then an increase in the number of passengers may not necessarily mean greater ancillary revenues.

An airport PPP must balance revenue and expenses. Government cannot simply assume that an airport will be profitable if there is a private operator. A robust feasibility study is essential.

Comparing the value of non-aeronautical revenues between bids can be tricky. It is easy to exaggerate the revenue potential for non-aeronautical services, and it may be difficult for the government to differentiate proposals at face value. El Loa ("Calama") airport in northern Chile was tendered as a PPP in January 2011, after the expiry of the first concession. The new concession included 8,100 m² of new terminal space for non-aeronautical services. The winning bid included an additional 2,000 m² of commercial space, arguing that additional revenues could be obtained and would increase profits. The larger commercial space played a major role in the highly profitable concession.[17]

Airport accounts can be managed on a single till, hybrid till or dual till system. Single till treats all airport revenues, aeronautical and non-aeronautical, in a single account. Dual till accounts for aeronautical and non-aeronautical revenues separately. The choice of till should be based on the proper management of the airport and the interests of consumers, and not just to maximise value for investors.

Box 3.2 Single till versus dual till regulation

Single till regulation means that a price cap is imposed on aeronautical charges, but taking into account profits earned in non-aeronautical activities. It is based on the principle of sharing airport costs between aeronautical charges and commercial revenues. This model effectively forces cross-subsidisation of aeronautical activities from any above normal non-aeronautical profits, and provides some incentives for cost minimisation in the provision of non-aeronautical and aeronautical activities, during the period of the price cap. Under a pure single till regulation, the more the airport makes on non-aeronautical activities (commercial), the less it has to charge the airlines and the passengers.

Under *dual till regulation*, aeronautical and non-aeronautical revenues and costs are registered in separate accounts. All aeronautical costs should be covered by aeronautical revenues. Therefore, regulation considers only the cost of production of aeronautical services when setting the fees and charges without taking into account the proceeds from non-aeronautical activities. Earnings from commercial activities are entirely for the airport.

There is an ongoing controversy between airlines and airports on the merits of single till vs. dual till. Airlines tend to advocate for the single till

---

[17] See "The Joy of Flying: Efficient airport PPP contracts" by Eduardo Engel, Ronald Fischer and Alexander Galetovic (2016).

arguing that a cross-subsidy between commercial activities and aeronautical operations is reasonable since it is the airlines that bring the business to the airports, justifying that the commercial revenues should be used to offset the cost base.

On the other hand, supporters of the dual till – in general the airports – argue that this approach yields economic efficiencies and provides the appropriate incentive for airport development.

3.7 Demand/traffic

Traffic volumes are the single most important driver of revenues and may have some impact on operational expenditures; they are the most important factor in risk assessment when considering commercial viability of an airport. Air transport activity involves two types of traffic: traffic generated by the airport catchment area, consisting of origin and destination passengers and cargo, and by connecting passengers and cargo transported between two other airports, brought into the airport by the airlines through their respective route networks.

Box 3.3 The myth of "Build it and they will come"

It is a common belief that "after the airport terminal expansion is completed, passenger traffic will increase". This belief is not necessarily related to capacity concerns, but rather wishful thinking. If aircraft movements during busy hours are reaching the operational capacity of the runway system, or not enough parking stands for the aircraft are available, traffic growth may be limited. Removing these constraints by expanding airside facilities typically increases traffic. However, the situation is different with the landside (mainly the terminal building). Terminal building congestion may decrease the level of service to passengers, but if there is a market, traffic will continue to grow. The truth is that neither business, nor tourist, nor visitor, friends and relatives (VFR) traffic is primarily motivated by the appearance of airport facilities. Traffic responds to market needs, irrespective of the appearance of the airport infrastructure.

A good example has been Cancún Airport, Mexico's second largest. Even before privatisation, traffic there grew at dramatic rates, even with passengers lining up in the sun or the rain at passport control. Many other touristic destinations are good examples, such as St Lucia Hewanorra, or Barbados Grantley Adams.

Investments in terminal buildings do not increase traffic, but are essential to provide a good level of service. Airports are gateways into countries or cities, a basis for pride and satisfaction. The level of comfort has a direct impact on the wellbeing of the general public.

48 *The business of airports*

Origin and destination (O&D) traffic is a function of diverse factors:

- *Economic context:* the national or regional economic situation is the single most important driver of passenger traffic growth. The local economy plays an important role in the generation of domestic traffic (business, tourism and "Visiting Friends and Relatives"[18]) as well as inbound international business travellers. The economic situation of other countries is also fundamental in generating tourism-driven international passengers. Destinations that rely heavily on tourism from specific markets are dependent on the economic situation of the source market. This is the case, for example, of San Jose and Liberia Airports in Costa Rica, dependent on the US and Canadian tourist markets, or the Southern Portuguese city of Faro, catering mainly to British tourists. The same applies for markets such as Mauritius, Seychelles or Maldives, and the impact of the European economic situation. Traffic can also be inversely affected by the economic context. For example, the devaluation of the Argentinean peso following the 2001 economic crisis generated a significant traffic growth at Buenos Aires airport as it became an affordable tourist destination.
- *Political context:* social unrest, political turmoil, terrorism and armed conflicts divert traffic, particularly tourism-related. After such incidents, the drop in leisure traffic is immediate and it may take considerable time to return to the previous traffic volumes. Egyptian airports, particularly those on the Red Sea, suffered significantly following the political turmoil of the spring of 2011. The political context may also have implications in the overall business risk for a PPP. Changes in the ruling government have resulted in termination of PPP contracts, as was the case in Budapest in 2001 and Male in 2012.
- *Geological/meteorological/ecological events:* natural disasters and extreme weather such as earthquakes, volcanoes, floods, tornadoes and storms may significantly disrupt air traffic, particularly tourism-driven. Ecological disasters, such as the Exxon Valdez oil spill in the Alaskan shores or the BP oil spill in the Gulf of Mexico, have affected airport traffic at tourist destinations. Some airports, particularly in the Caribbean, take into consideration probable disruption of activities during the summer season of hurricanes, accounting for these effects in air traffic forecasts. The eruptions of the Eyjafjallajökull volcano in Iceland in 2010 caused major traffic disruptions across the northern Atlantic routes that lasted almost a week.
- *Specific one-time events:* a football World Cup, Olympic Games or other special event may cause significant distortions of traffic patterns. During the Brazil Football World Cup, while there was a dramatic increase of inbound tourists, Brazilians significantly reduced their overseas and domestic travel. A derived effect from these events is an improvement in tourism capacity and infrastructure and destination awareness among foreign tourists.

---

18 Known as "VFR" (and not to be confused with "visual flight rules"!).

- *Regulatory issues:* specific regulatory issues can profoundly affect traffic volumes. Many airports have experienced traffic losses following enactment of new regulations that restrict airport operations. For example, after the terrorist attacks of September 2001, non-US citizens connecting through US airports were required to have a valid visa in their passports. As a result, a significant share of traffic that used to connect at Miami between Central America and South America, or even Europe, moved to other hubs in the region. Panama City Tocumen Airport benefited from this situation, as Copa Airlines succeeded in providing an efficient network for intra-American connectivity.
  - Another example is Mozambique: after a visa waiver agreement with South Africa in 2005, South African tourist traffic at Mozambican airports increased significantly. Relaxed travel restrictions for Chinese citizens has created a significant impact on traffic demand across the world.
  - Imposing night operational curfews have reduced the overall capacity of large European hubs. Any change in these types of regulations can significantly affect overall traffic, particularly at constrained airports. Charter flights, that preferred to operate at off-peak periods such as night time, might be particularly affected by night curfews.
- *Technical issues:* traffic can be affected by issues of a strictly technical nature. Topography, operational limitations such as runway length, obstacles to airport approach or climb, winds, no-fly zones, could constrain traffic growth. Technical issues that limit current or future operations should be assessed and considered. A significant number of domestic flights, accounting for as many as three million passengers, were transferred from São Paulo Congonhas Airport to São Paulo Guarulhos Airport after a 2008 crash, attributed mainly to infrastructure deficiencies. For years, at Kathmandu Tribhuvan, the lack of night landing capabilities has limited the potential to offer a direct link to the significant Japanese tourist market that cannot access Nepal in one day and must spend a night in Seoul. Siem Reap airport in Cambodia faces operational restrictions that hamper traffic growth from specific tourist markets due to limited runway length and the no-fly zone over Angkor Wat, a historic world heritage site. These limitations in Cambodia restrict aircraft size and takeoff payload to the extent that it is unprofitable for airlines to operate direct flights from Korea and Japan, two of the most important sources of tourists to Siem Reap. The Tegucigalpa Toncontin Airport in Honduras has a complicated flight approach between mountains and valleys, which has been a concern to successive government administrations. The airport altitude, topological features, and runway length limits airport growth and restricts the type of equipment airlines can use, increasing pressure to find a longer-term alternative solution. Because of the extremely complicated approach at Paro International Airport in Bhutan, only pilots from the two local airlines are certified to land, limiting traffic.
- *Airline strategies:* when O&D passengers have different alternatives to choose from, airline strategies can impact on demand for a specific airport. An example

was the passenger traffic at Porto Francisco Sá Carneiro Airport, in Portugal. Spanish Galicians travelling to Caracas, home of their largest diaspora, are dispersed among three airports (La Coruna, Vigo and Santiago). Because Iberia's network only offers international services from Madrid, Galicians used to travel by surface to Portugal's city of Porto to board TAP direct flights to Venezuela. Porto would not have served this large population of Spaniards had Iberia offered direct international services from the Galician cities.

- Connecting traffic is a function of:
  - *Airline networks:* passengers connecting between flights are funnelled through airports that are neither their origin nor their destination, but their transfer/transit point between flights. Changes in the way airlines structure their networks result in changes in passenger volumes. For example, Lufthansa's strategy to diversify connecting traffic from Frankfurt to Munich, after the opening of Terminal 2, resulted in traffic volume increases at Munich. Another example is the significant loss of long-haul O&D and connecting traffic in Brussels airport after the collapse of Sabena, and the availability of efficient rail links to other airports. Cincinnati/Northern Kentucky International Airport lost an aggregated 51% of its traffic between 2008 and 2011 after Delta merged with Northwest and reduced flight capacity from one of its main hubs.
  - *Aviation policy regulations:* government policy decisions can affect airport traffic. The progressive liberalisation of transit between countries contributed to the development of traffic at many airports. For example, the open skies agreement between the US and the Dominican Republic has been a key factor in the significant development of tourism on the Caribbean island. Similarly, the reforms implemented by the government of Georgia contributed to the country's development, doubling the passenger traffic at Tbilisi airport in just four years.
- *Competing modes:* competing transport modes such as high-speed trains or highways can reduce passenger traffic for short and medium distances. The opposite has also been true. Land transport was dangerous in Colombia for many years, contributing to the development of one of the most comprehensive air transport networks in Latin America. High speed trains may have some negative impact on particular city pair markets. The introduction of the AVE high speed train between Madrid and Seville increased rail market share on that route from 33 to 84%.[19]

Traffic demand risk assessment requires a comprehensive market analysis and demand forecast. Analysis should consider drivers of future traffic, including economic development (economic growth, tourism development, accommodation capacity), political context (political risk, social unrest, war, terrorism), airline

---

19 See Rodriguez et al. (2009), *The Geography of Transport Systems* (based on data from the International Union of Railways).

plans for any change in route networks, fleet renewal, new entrants, airline mergers or consolidation, competing airports, etc. Traffic forecasts should combine statistical analysis with robust market knowledge.

Demand risk for non-aviation related activities, mainly real estate not connected to airport activity, should be estimated based on specific drivers related to each activity (industrial, retail, office space, housing, and so forth). The USD 5 billion Brisbane Airport Link toll road went into insolvency in 2013. Traffic levels achieved only 40% of the forecast 135,000 vehicles per day.[20]

Box 3.4 Demand for air traffic movements

The demand for aircraft movements (ATMs) is a derived effect from passenger demand, a function of the number of passengers that will be arriving per flight. This number is a function of:

Aircraft size: the fleet composition of an airport is a direct reflection of the role of the airport.

- Intercontinental hubs receive large, wide-body aircraft flying long-haul routes, and feed the traffic through short- and medium-haul spokes, resulting in a large number of medium-size aircraft. Such is the case for example of Frankfurt Airport, or Paris Charles de Gaulle, where long haul traffic received on Boeings 747-8, Airbuses 380, Boeings 777 and Airbuses 330, or 787s and A350s, are then distributed on narrow bodies like the B737 or the A320 family. Other intercontinental hubs, like Dubai International, connect long-haul flights, resulting in a greater proportion of wide-bodied aircraft
- Country gateway airports provide for domestic distribution, and therefore host a greater proportion of narrow-bodied aircraft
- Regional airports serve primarily small, narrow-bodied (up to 120-seaters) and turboprop craft, with a lesser proportion of narrow bodies connecting to larger airports
- General aviation airports are dedicated to small business or recreational craft, with some minor transport aircraft like 12 to 19-seater turboprops operating very thin markets.

The average airplane size has grown steadily at about 1% per year for the last 20 years.[21] Growth in the market is expected to continue through the proliferation of non-stop markets and increases in frequency, requiring more narrow bodies on point to point markets, not necessarily through main hubs. The largest proportion of aircraft orders from the two largest

---

20 This data is again from IATA's 2018 guidance booklet on airport ownership and regulation.
21 This data is from "Boeing Commercial Outlook: Commercial Market Forecasting", May 2019.

manufacturers, Boeing and Airbus, are chiefly narrow bodies. While Airbus has announced the termination of the A380 program, Boeing's production of its 747-8 is almost exclusively dedicated to cargo.

Load factor: the proportion of the aircraft seats occupied is the load factor. Lower fares and more efficient revenue management techniques help airlines achieve higher load factors. LCCs tend to achieve higher load factors than legacy carriers, particularly through their selling techniques as well as a higher proportion of point to point operations, which are less dependent on passengers arriving from a connecting flight.

3.8 Other issues

A number of other key issues will impact the project financials.

Environmental issues can deter airport expansion. Mexico City International Airport (Benito Juarez) has been surrounded by controversy for over a decade because of environmental opposition to build into the Texcoco Lake, risking potential traffic growth. These kinds of environmental risks can impede potential for implementing a successful PPP.

Administrative issues. Changes to airport management can be complicated due to restrictions to workforce layoffs, changes to remunerations levels or use of unionised employees, service contract interruptions, grandfathered rights of vendors or service providers, among others. All must be evaluated for achievability and timing.

Social issues include any developments that may affect neighbouring populations. In some areas of the world, such as Madagascar, dwelling encroachments on airport property are common and can be a safety hazard or restrict facilities operations or expansion. Relocating settlements can pose political risks for the Government, not to mention the associated economic and financial costs. In the UK, London Heathrow Airport may lose its status as the world's largest international airport if surrounding communities continue blocking the construction of its third runway. Frankfurt Airport faced a critical moment upon the opening of the fourth runway and the political pressure from surrounding communities imposing a ban on night flights. These risks must be evaluated and factored into anticipated airport activity.

Institutional issues. The necessity to maintain equilibrium between different stakeholders may pose additional risks. Many airport facilities are used for civilian and military purposes; the relationship is not always simple and expansion of the operations can be limited by lack of agreement between the parties. For example, the Bogotá Eldorado airport master plan had to consider the presence of the military facility right across from the old main terminal, occupying premium space, for the expansion of the airport. Issues of this sort have been a major disruption at airports jointly used by commercial and military operations, forcing the commercial flights out to new greenfield facilities, as in the cases of Natal (Brazil) and Goa Dabolim Airport.

3.9 Government support

Governments can and should provide a number of financial mechanisms to support and enable PPP.[22] Some airport PPP projects, in particular small or medium airports, require some form of government technical or financial support. Efficient financing of PPP projects can involve the use of government support, to ensure that the government bears risks which it can manage better than private investors and to supplement projects which are economically but not financially viable, i.e. where an airport projects has large public externalities, some level of direct financial support from the government may be appropriate. Also, local financial markets may not be able to provide the financial products (in particular long term, fixed interest debt) needed for PPP, even though such products would benefit the entire financial market. The government can do much to resolve these issues, providing funded or contingent products, or by creating entities that provide some form of financial support necessary for PPP to flourish.

Each project is likely to require tailor-made support, but the individual instruments to be used should be carefully designed to provide the perceived predictability that the private investor needs and flexibility the government needs. When considering government support, the government will need to consider carefully:

- Which projects to support
- How much support to provide
- The terms of such support; and
- How to ensure that support is properly managed, e.g. transparency and proper governance.

In particular, government support can create conflicts of interest or misaligned incentives, as the government will be playing different roles on different sides of the transaction; e.g. it may be contracting authority and shareholder at the same time. Government support therefore needs to be designed, implemented and regulated carefully.

These mechanisms are particularly useful where the project does not on its own merit achieve bankability, financial viability or is otherwise subject to specific risks that private investors or lenders are not well placed to manage. In

---

22 For more, see Delmon, *Private Sector Investment in Infrastructure* (2009).

developing countries where private finance is most needed, these constraints may necessitate more government support than would be required in more developed countries.

The decision to provide government support should be finalised and announced well before the project bid date, to improve the private investor's appetite, increase the number of bids and reduce project costs for the contracting authority. The commercial and pricing benefits from government support packages are effective only if the support is announced in advance of the competition and if it is well designed. Allocation of government support after the bid date will deprive the contracting authority of most of these benefits.

3.9.1 Funded products

Funded support involves the government committing direct financial support to a project, such as:

- Grant contributions in cash or in-kind (e.g. to defray construction costs, to procure land, to provide assets, to compensate for bid costs or to support major maintenance)
- Waiving fees, costs and other payments which would otherwise have to be paid by the project company to a public sector entity (e.g. authorising tax holidays or a waiver of tax liability)
- Providing financing for the project in the form of loans (including mezzanine debt) or equity investment; and
- Funding shadow fees and charges and topping up fees and charges to be paid by some or all consumers (for example, local airlines may not have paid fees and charges previously, and government may choose to subsidise these costs during a transition period to reduce the burden on local airlines and to help manage the airport operator's collection risk).

These mechanisms can be used in combination, and can be more or less targeted.

3.9.2 Contingent products

The government may choose to provide contingent mechanisms, i.e. where the government is not providing funding, but is instead taking on certain contingent liabilities, for example providing:

- Guarantees, including guarantees of debt, exchange rates, convertibility of local currency, collection of fees and charges, the level of demand for specific services, termination compensation, etc.
- Indemnities, e.g. against non-payment by public entities (e.g. where the national flag carrier is a major user of airport services but is not credit-worthy, for revenue shortfall, or cost overruns

- Insurance
- Hedging of project risk, e.g. adverse weather, currency exchange rates or interest rates; or
- Contingent debt, such as take-out financing (where the project can only obtain short tenor debt, the government promises to make debt available at a given interest rate at a certain date in the future) or revenue support (where the government promises to lend money to the project company to make up for revenue shortfalls, enough to satisfy debt-service obligations).

The government will want to manage the provision of government support, and in particular any contingent liabilities created through such support mechanisms. Governments seek a balance between (i) supporting private infrastructure investment and (ii) fiscal prudence.[23] Striking this balance will help the government make careful decisions about when to provide public-money support and manage the government liabilities that arise from such public-money support, while still being aggressive in encouraging infrastructure investment. Government assessment of projects receiving such support is doubly important given the tendency of lenders to be less vigilant in their due diligence when government support is available, since this reduces lender risk and exposure.

Governments actively managing fiscal risk exposure face challenges associated with gathering information, creating opportunities for dialogue, analysing the available information, setting government policy and creating and enforcing appropriate incentives for those involved.[24] Given the complexity of these tasks, it is becoming more popular for governments, and in particular ministries of finance, to create specialist teams to manage fiscal risk arising from contingent liabilities, in particular those associated with PPP across all sectors. This is often achieved through debt management departments, which are already responsible for risk analysis and management.[25] The government may also consider creating a separate fund to provide guarantees, allowing the government to regulate this function better and ring fence the associated government liabilities.

3.9.3 Project development funds

Airport PPP projects require upfront investment in project preparation, which often meets with strong resistance from government budgeting and expenditure control authorities. However, quality advisory services are key to successful PPP development, and can save millions in the long-run.[26] Therefore, funding,

---

23 For further discussion of this issue, see Irwin, *Government Guarantees: Allocating and Valuing Risk in Privately Financed Infrastructure Projects* (2007).
24 Brixi, Budina and Irwin, "Managing Fiscal Risk in Public Private Partnerships" (2006).
25 Different structures may be needed to track and manage risks created by central and local governments.
26 HM Treasury (UK), "PFI: Strengthening Long-term Partnerships" (March 2006).

budgeting and expenditure mechanisms for project development are important to a successful PPP program, enabling and encouraging government agencies to spend the amounts needed for high quality project development.

The government may wish to develop a more or less independent project development fund (PDF), designed to provide funding to contracting authorities for the cost of advisers and other project development requirements, and focused on airport PPP or more generally for PPP project development. The PDF may be involved in the standardisation of methodology or documentation, its dissemination and monitoring of the implementation of good practices. It should provide support for the early phases of project selection, feasibility studies and design of the financial and commercial structure for the project, through to financial close and possibly thereafter, to ensure a properly implemented project. The PDF may provide grant funding, require reimbursement (for example, through a fee charged to the successful bidder at financial close) with or without interest, or obtain some other form of compensation (for example, an equity interest in the project), or some combination thereof, to create a revolving fund. The compensation mechanisms can be used to incentivise the PDF to support certain types of projects.

Box 3.5 Factors minimising government contributions in concessions/BOT

1. *Low capital requirement*: capital investment is financed by debt and equity. More investment results in more debt repayment, which may require government support. Phased investment or less flamboyant airport visual design may reduce such investment needs
2. *High revenue potential*: airports with sufficient traffic can generate more revenue if better managed. One option is to increase fees and charges if it does not impact the competitiveness of the airport. Another option is to allow the private operator to maximise the commercial potential, including retailing (duty free and duty paid), food and beverage and other commercial activities
3. *Improved financial terms*: government can improve indirectly financial terms by reducing the overall project risk. Revenue in foreign currency for example reduces the foreign exchange risk and encourages borrowing in foreign currency, usually associated with longer maturity and lower interest rates
4. *Improving the overall business proposition*: government may decide to include other components in the project which could add value, such as free trade zones, industrial parks and areas for real estate development (not related to the airport). In addition, the project can be offered with a longer than usual period, to assure a better return on investment for the equity holders.

4 Airport planning

A goal without a plan is just a wish.
Attributed to Antoine de Saint-Exupéry (1900–1944),
French writer, poet and aviator

Planning is key. For an asset as strategic as an airport to deliver the benefit to the economy and to investors that it is able, it must fit within a clear plan for development, for transport and for airports.

The planning process starts with strategy, to understand the direction of the sector, the direction of government and how air transport fits into that strategy. The

58 Airport planning

strategy must respond to demand, the location of traffic, and how much the airports need to grow to respond to demand. This involves good, disaggregated data on traffic today and a robust forecast of traffic in the future. The plan also needs to respond to other operational and other issues that reflect practical constraints to the plan.

The government plan then needs to be focused in a plan specific to the airport in question, that feeds from the national plan, but responds to the specific context of the airport in question – investment needs, safety, security and service standards.

4.1 Strategy

When planning or developing an airport, policy makers must face a number of strategic considerations:

- National pride: many airports are designed as national monuments and country gateways, arousing strong emotions in the government and the travelling public (e.g. Munich Terminal 2, Washington Dulles, Bangkok Suvarnabhumi, Dubai World Central)
- Award winning: some airports are intended to be the international champions of level of service, the preferred choice for all travellers, regardless of the cost of construction (e.g. Singapore Changi, Doha Hamad International)
- Tourism development: some airports are developed as key instruments for tourism development to ensure a progressive increase in incoming travellers (e.g. Bali Ngurah Rai, Punta Cana, Nosy Be Fascene)
- World largest international hub: airports that aim to be the largest international hubs in the world (e.g. London Heathrow, Dubai World Central)
- Hub for a national carrier: an airport may be constructed to attract a specific airline to base its operations at the airport, possibly designed according to the needs of the carrier (e.g. Denver International, Panama Tocumen, Madrid Barajas T4S)
- Profitable: some airports are designed with materials and facilities that provide a good service while achieving the highest return to its shareholders (e.g. Lima Jorge Chávez Airport, Athens Eleftherios Venizelos)
- Easy access: some airports are maintained at their current location for ease of access (e.g. Washington National Reagan, Lisbon Portela, Buenos Aires Aeroparque Jorge Newberry)
- Environmental leader: an airport might be by design environmentally sustainable, a leader in environmental and energy design (e.g. Stockholm Arlanda, Boston Logan)
- Technology leader: airports that seek to be at the cutting edge of technological innovation as their core objective (e.g. Singapore Changi, San Francisco)
- Social considerations: the airport might respond to social necessity, to connect remote communities or to provide easier access to air transportation, for example, in archipelagic countries (e.g. São Filipe Airport, on the island of Fogo in Cape Verde)

- National security considerations: an airport might be developed as an alternative facility in case of an attack or other disaster at the main airport, to ensure continuous connectivity. For example, smaller airports may have an excessively long runway and large apron to receive aircraft from other airports (e.g. Timna Ramon International Airport, Israel).

There are various examples of large airports designed from scratch (as greenfield projects) that lacked strategic analysis and therefore resulted in complete failures. Ciudad Real in Spain is an example of a greenfield airport that cost over one billion euros and was operated for only three years before shutting down.[1]

4.2 Associated infrastructure

Associated infrastructure must also form part of the government strategy. For example, access to airport, by road and rail, is important to ensure demand.

In 2005, the collapse of a major bridge in the highway connecting Maiquetía Airport with downtown Caracas, in Venezuela, resulted in severe disruptions in accessing the airport; the only alternative, winding roads through the mountains, turned the usual 25-minute ride into a two-hour trek. It took the government over a year to reconstruct the bridge, seriously affecting the traffic demand.

Distance from the city and access time was seriously underestimated when the Government of Canada planned Mirabel Airport, to replace the more proximate airport in Dorval. The travelling public never accepted the new location and the airlines refused to move. As a result, the project become one of the biggest failures in airport planning, crowned by the demolition of a new passenger terminal building to make a race track.

Tokyo Narita International was built for international traffic, at about 67 kilometres from the city centre, replacing Haneda Airport, which is just 17

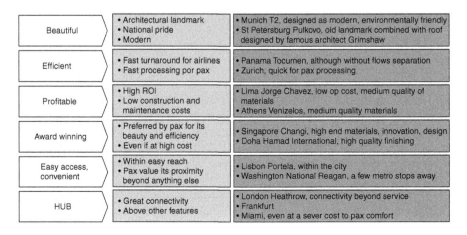

Figure 4.1 Airport strategies

1 It was only used by the Spanish director Almodovar to film a movie.

kilometres away from Tokyo. Haneda was restricted to domestic flights and for those international routes that are shorter than the longest domestic segment. In 2008 the authorities allowed night flights and, following a new runway and a new terminal building, the airport lifted all regulations in 2013. The effect was dramatic: Haneda's traffic grew at an average pace of 27.7% against Narita's 2.7%. The easier access of Haneda Airport has been fundamental in the passenger preference over Narita, representing savings in cost and time.

Severe traffic jams on the road to Nairobi Jomo Kenyatta International contributed to the development of local operations out of Wilson.

Interisland traffic connecting St Lucia with neighbouring islands is facilitated from George F. L. Charles Airport located next to the island capital city centre of Castries. The over 90-minute drive needed to reach Hewanorra International at the southern tip of the island would discourage a significant amount of the interisland traffic, because of the time and cost involved.

Attempting to combine the construction of the airport together with the road link under the same PPP arrangement is not recommended. Airport and road developers represent different operator profiles, making the two projects difficult to coordinate under one single PPP.

The coordination between the completion of a new airport and its link access should be managed carefully. Quito Mariscal Sucre airport, a greenfield PPP airport developed about 37 kilometres from the city centre, was inaugurated in February 2013. However, a connecting road of 23 kilometres was not completed until 17 months later, creating serious disruptions in access.

Athens Venizelos International Airport, completed in 2001, is located slightly over 30 kilometres from the city centre. The airport was inaugurated ahead of the completion of the Attiki Odos motorway, creating some access disruptions. The Government of Greece engaged Hochtief under a PPP to develop the motorway, facilitated by the European Investment Bank, which financed both projects. The development of the road, including timing and standards, was set out in the airport concession contract, including hefty penalty clauses for delay.[2]

4.3 Traffic forecasting

Traffic is one of the most important drivers for airport planning. This said, an airport does not need heavy traffic to be strategically important. Location, special facilities and other characteristics can make even the smallest airport strategic, but for the great majority of the airports, passengers, aircraft movements and cargo are the three main drivers of strategic relevance. Understanding the dynamics of traffic demand and forecasting them accurately are key for planning.

Air transport activity is derived from two types of traffic: traffic generated by the airport catchment area, consisting of origin and destination passengers and cargo; and connecting traffic travelling between two other airports, brought into

2 See the UCL Bartlett School of Planning's Project Profile on Attiki Odos, available online.

Airport planning 61

the airport by the airlines through their respective route networks. These issues are discussed in more detail in Chapter 3, but in summary:

Traffic is a function of diverse factors:

- *Economic context*: the national or regional economic situation is the single most important driver of passenger traffic growth
- *Political context*: social unrest, political turmoil, terrorism and armed conflicts risks divert traffic, particularly tourism-related
- *Geological/ecological events*: natural disasters and extreme weather such as earthquakes, volcanoes, floods, tornadoes and storms may significantly disrupt air traffic, particularly tourism-driven
- *Regulatory issues*: specific regulatory issues can impact service levels, fees and charges and other requirements of airports that can have a direct impact on planning
- *Technical issues*: traffic can also be affected by issues of strictly technical nature, including the location and context of the airport (space, location) and access to latest technology. Topography, operational limitations such as runway length, obstacles to the airport, winds, no-fly zones, or night curfew restrictions, for example, could constrain traffic growth
- *Airspace restrictions*: some air transport markets face greater challenges in the air than at the airport. Inefficient air navigation service providers, or limited access corridors because of bottlenecks in neighbouring countries, can restrict the capacity at a given airport
- *Airline strategies*: when airline networks provide O&D passengers with alternatives, this will affect demand for a specific airport
- *Competing modes of transport*: competing modes of travel such as high-speed trains or highways can reduce passenger traffic for short and medium distances
- *Environmental issues* can deter airport expansion, where needed land is environmentally sensitive; consultation processes with surrounding communities are particularly sensitive
- *Administrative issues*: implementing changes in airport operation may be particularly difficult, for example restrictions to workforce layoffs, changes to remunerations levels or use of unionised employees, service contract interruptions, grandfathered rights of vendors or service providers, among others
- *Social issues* include impact on neighbouring populations, such as dwelling encroachments on airport property and the need to relocate settlements to access needed land
- *Institutional issues*. Inter-institutional equilibrium may be difficult to achieve, for example where an airport is used for civilian and military purposes.

Analysis should consider drivers of future traffic, including economic development (economic growth, tourism development, accommodation capacity), political context (regional alliances, political risk, social unrest, war), airline plans for changes in route networks, fleet renewal, new entrants, airlines mergers or consolidation, competing airports, etc. Traffic forecasts should combine statistical analysis with robust market knowledge.

Demand risk for non-aviation related activity, mainly real estate not connected to the airport activity, should be estimated based on specific drivers related to each activity (industrial, retail, office space, housing, and so forth).

4.4 Master planning

The national or sector-wide plan needs to be applied to an individual airport, to understand how the airport will develop physically over time, to meet projected demand, to comply with service level agreements (SLA) and to achieve investment requirements. A master plan is the physical expression of the government's strategic plan and the airport business plan.

Typically, master plans are developed with a five-year horizon, and at five-year intervals, planning for the transformation that is expected over the next 30 years. A comprehensive master plan should start from a traffic demand projection, the associated investments to meet such demand, and the projected financials that incorporate the business expectation and the need for capital expenditures.

The importance of reviewing the master plan at periods no longer than five years is to adapt the projections of traffic, investments and financials according to changes in the market, user needs, adaptation to new technologies, environmental regulations and changes in strategies. A master plan should not be more than a referential plan of how the airport will evolve, but never at a detailed engineering level.

Box 4.1 Elements of an airport master plan

The elements of an airport master plan vary with each airport's situation and may include the following:

1. Inventory of all relevant airport infrastructure, land uses, operational activities, and development issues.
2. Forecasts of aeronautical demand on airport infrastructure for subsequent 5-, 10-, 20- and 25-year periods.
3. Determination of future airport requirements and alternative concepts for demand satisfaction. This step involves comparison of present capabilities with forecast demand at five-year intervals, and the development of concepts to best satisfy different infrastructure needs.
4. Site location and condition of new, relocated, or expanded infrastructure.
5. Environmental analysis and procedural compliance.
6. Airport plan drawings (a series of large-scale drawings of overall airport layout, terminal area plan, land use plan, surface transportation access, and the airspace plan).

Source: Handbook of Transportation Engineering

Table 4.1 Alternative airport roles

Airport role	Features	Examples
International hub	• Connects intercontinental long haul flights with other intercontinental long/medium haul flights	• London Heathrow, Dubai International, Frankfurt Flughafen
Regional hub	• Connects international long haul flights with regional destinations (medium and short haul), serving as a regional gateway • Also has an important origin and destination traffic	• Frankfurt Flughafen, Paris Charles de Gaulle, London Heathrow, Amsterdam Schiphol, Miami International, Lima Jorge Chávez, Hong Kong, Narita Tokyo, Singapore Changi, Addis Ababa Bole, Johannesburg O.R. Tambo
Country gateway	• Serves as an entry point from international routes into the country • Base for the main local airline	• Madrid Barajas, Paris Charles de Gaulle, Chicago O'Hare, Toronto Pearson, Mexico Benito Juarez, Casablanca Mohammed V, Bogotá El Dorado, Moscow Sheremetyevo, Antananarivo Ivato, Johannesburg O.R. Tambo
International terminal airport	• Because of its geographical location it is mainly origin and destination traffic with few connections • Without relevant domestic market from this particular airport • No main carrier hubbing at the airport	• Barcelona El Prat, Athens Venizelos, Buenos Aires Ezeiza, Sao Paulo Guarulhos, Brisbane, Tel Aviv Ben Gurion, Lagos Murtala Muhammed
Low Cost Carrier base	• Mostly a secondary airport at a main city where low cost carriers enjoy special operational and commercial conditions to base their operations	• Dallas Love Field, Kyiv Sikorsky (Zhuliany), Charleroi, Bergamo, London Stansted, Bangkok Don Muang, Buenos Aires El Palomar, Frankfurt Hahn, Manila Clark, Tuzla
Leisure oriented traffic	• Oriented to leisure traffic destinations, regionally or long haul • Serves main charter airlines	• London Gatwick, Düsseldorf, Paris Orly
Cargo airports	• Airports with a focus on freight operations	• Hong Kong, Memphis, Louisville, Wilmington, Shanghai, Anchorage, Incheon, Sao Paulo Viracopos, Liege, Leipzig, Dubai World Central

Source: authors

64 Airport planning

The US Federal Aviation Administration describes the goal and objectives of an airport master plan:

a. Airport master plans are prepared to support the modernization or expansion of existing airports or the creation of a new airport. The master plan is the sponsor's strategy for the development of the airport.
b. The goal of a master plan is to provide the framework needed to guide future airport development that will cost-effectively satisfy aviation demand, while considering potential environmental and socioeconomic impacts. The FAA strongly encourages that planners consider the possible environmental and socioeconomic costs associated with alternative development concepts, and the possible means of avoiding, minimizing, or mitigating impacts to sensitive resources at the appropriate level of detail for facilities planning.
c. Each master plan should meet the following objectives:
   1) Document the issues that the proposed development will address.
   2) Justify the proposed development through the technical, economic, and environmental investigation of concepts and alternatives.
   3) Provide an effective graphic presentation of the development of the airport and anticipated land uses in the vicinity of the airport.
   4) Establish a realistic schedule for the implementation of the development proposed in the plan, particularly the short-term capital improvement program.
   5) Propose an achievable financial plan to support the implementation schedule.
   6) Provide sufficient project definition and detail for subsequent environmental evaluations that may be required before the project is approved.
   7) Present a plan that adequately addresses the issues and satisfies local, state, and Federal regulations.
   8) Document policies and future aeronautical demand to support municipal or local deliberations on spending, debt, land use controls, and other policies necessary to preserve the integrity of the airport and its surroundings.
   9) Set the stage and establish the framework for a continuing planning process. Such a process should monitor key conditions and permit changes in plan recommendations as required.[3]

ICAO defines the following process for developing a master plan, shown in Figure 4.2.

---

3 From the US Department of Transport/FAA's 2005 Advisory Circular on Airport Master Plans, AC 150/5070-6B.

Figure 4.2 The process of master planning
Source: Inspired by David Stewart, ICAO Airport Planning Seminar for the SAM Region, Lima, Peru, September 2018, from the IATA *Airport Development Reference Manual*

In summary, a master plan provides the analytical framework for a project's feasibility and constitutes an important phase of any airport development plan.

The master plan will generate the capital expenditure program and an associated cash flow from financing activities that will be incorporated in the airport projected cash flow.

As explained in Chapter 3, investments will be a function of the required infrastructure needs to comply with standards of safety and security, to accommodate the current and future demand, and to carry out major repairs.[4]

---

4 Please see Chapter 3.5.2, where we define capex in more detail.

4.5 Consultation

During the developing of different options and before selection of the preferred approach, consultation with key stakeholders is critical. There is no more important partner to involve than the airlines. New facilities must be consistent with their operations. Locally based airlines may be the most critical partners, where they pose the largest stress on the facilities. Their involvement at the planning stage can provide early warning of possible defects or even designs that could affect negatively the operation of aircraft.

There are numerous cases where airport planners have overlooked the implications of a particular design for an airline. Among the many examples, the design of the new Panama Tocumen International Airport omitted a proper dialogue with their main airline, Copa Airlines. Branded by Copa as "the hub of the Americas", Tocumen Airport is one of the two largest hubs in Latin America. The recently opened Terminal 2 inaugurated in April 2019, lacks the sufficient number of gates to accommodate all Copa's operations during a hub bank.[5] Copa will still depend on gates located at the old terminal, increasing the minimum connecting times for their transfer passengers to onward flights.

Another key stakeholder is the local community. Consultation processes help the airport understand local concerns, including environmental issues with respect to noise and traffic. The UK's Department of Transport carried out an extremely thorough consultation process for the construction of a third runway at London Heathrow. Failure to consult with the local community has resulted in delays and even failures of airport projects.

---

5 A "hub bank" is the period in the day when flights arrive at the airport and passengers change flights for outward connections.

5   Air transport regulation

> Those are my principles, and if you don't like them... well, I have others.
> Groucho Marx, aka Julius Henry Marx (1890–1977),
> American comedian and actor

This chapter outlines the key drivers of airport regulation, starting with the international conventions and organisations that form the basis for the regulatory framework: the International Civil Aviation Organization (ICAO) (section 5.1) and the International Air Transport Association (IATA) (5.2); followed by a discussion of the fundamental elements of the regulatory framework: Air Service Agreements (5.3), slot allocation (5.4), and economic regulation (5.5).

68 *Air transport regulation*

5.1 The International Civil Aviation Organization (ICAO)

In 1944, the U.S. government extended an invitation to 55 countries to attend an International Civil Aviation Conference in Chicago. Known then and today more commonly as the "Chicago Convention", this landmark agreement laid the foundation for the standards and procedures to achieve peaceful global air navigation. It set out as its prime objective the development of international civil aviation "in a safe and orderly manner", and such that air transport services would be established "on the basis of equality of opportunity and operated soundly and economically".[1]

The Chicago Convention also proposed the establishment of a specialised International Civil Aviation Organization (ICAO), to organise and support intensive international co-operation to deliver a global air transport network. Its core mandate was to help countries achieve uniformity in civil aviation regulations, standards, procedures, and organisation.

In 1947, ICAO became a United Nations (UN) specialised agency. While ICAO remained an independent and autonomous agency, its acquisition of constituency status in the UN Organization benefited many of its Member States in the years which followed, mainly through the United Nations Program of Technical Assistance.

Currently ICAO works with the Convention's 193 Member States and industry groups to reach consensus on international civil aviation Standards and Recommended Practices (SARPs) and policies in support of a safe, efficient, secure, economically sustainable and environmentally responsible civil aviation sector. These SARPs and policies are used by ICAO Member States to ensure that their local civil aviation operations and regulations conform to global norms.

The SARPs are contained in 19 Annexes to the Chicago Convention, as follows:

- Annex 1 – Personnel Licensing
- Annex 2 – Rules of the Air
- Annex 3 – Meteorological Service for International Air Navigation
- Annex 4 – Aeronautical Charts
- Annex 5 – Units of Measurement to be Used in Air and Ground Operations
- Annex 6 – Operation of Aircraft
- Annex 7 – Aircraft Nationality and Registration Marks
- Annex 8 – Airworthiness of Aircraft
- Annex 9 – Facilitation
- Annex 10 – Aeronautical Telecommunications
- Annex 11 – Air Traffic Services
- Annex 12 – Search and Rescue
- Annex 13 – Aircraft Accident and Incident Investigation

---

1 Quotes from "The History of ICAO and the Chicago Convention", available on the ICAO's website.

- Annex 14 – Aerodromes
- Annex 15 – Aeronautical Information Services
- Annex 16 – Environmental Protection
- Annex 17 – Security: Safeguarding International Civil Aviation against Acts of Unlawful Interference
- Annex 18 – The Safe Transport of Dangerous Goods by Air
- Annex 19 – Safety Management

In addition to its core work resolving consensus-driven international SARPs and policies among its Member States and industry, ICAO also:

- Coordinates assistance and capacity building for Member States in support of numerous aviation development objectives
- Produces global plans to coordinate multilateral strategic progress for safety and air navigation
- Monitors and reports on numerous air transport sector performance metrics; and
- Audits States' civil aviation oversight capabilities in the areas of safety and security (Universal Safety Oversight Audit Programme – USOAP).

5.1.1 ICAO objectives

ICAO standards are based on a number of objectives:

Safety: ICAO works in close collaboration with institutions, governments and agencies across the global air transport industry to further improve aviation's safety performance while maintaining a high level of capacity and efficiency. This is achieved through:

- The development of global strategies contained in the Global Aviation Safety Plan and the Global Air Navigation Plan
- The development and maintenance of SARPs applicable to international civil aviation activities which are contained in 19 Annexes
- The monitoring of safety and air navigation trends. ICAO audits the implementation of its SARPs through its USOAP. It has also developed tools to collect and analyse an array of aviation data which allows it to identify existing and emerging risks
- The implementation of targeted safety and air navigation programs to address deficiencies
- An effective response to disruption of the aviation system caused by natural disasters, conflicts or other causes.

Economic development: Fostering an economically viable civil aviation system (airlines, airports, air navigation services providers, etc.) and enhancing its economic efficiency and transparency while facilitating access to funding for aviation

70 *Air transport regulation*

infrastructure and other investment needs, technology transfer and capacity building to support the growth of air transport for the benefit of all stakeholders.
Environmental protection: ICAO has developed a range of standards, policies and guidance material to address aircraft noise and emissions embracing technological improvements, operating procedures, proper organisation of air traffic, appropriate airport and land-use planning, and the use of market-based options. In 2004, ICAO adopted three major environmental goals:[2]

- Limit or reduce the number of people affected by significant aircraft noise
- Limit or reduce the impact of aviation emissions on local air quality; and
- Limit or reduce the impact of aviation greenhouse gas emissions on the global climate.

5.1.2 Airport development and planning

Three particular annexes of the ICAO Convention are closely related to airport development and planning.

Aerodromes (Annex 14) covers all aspects planning of airports and heliports, detailing aerodrome data, physical characteristics (such as runways, taxiways, aprons, etc.), obstacle restrictions, visual aids (for navigation, for denoting obstacles and restricted used areas), electrical systems, operational services (such as rescue and firefighting, emergency planning, etc.), and aerodrome maintenance.

This Annex is one of the most rapidly changing, due to the constant evolution of technologies with respect to airport operations, models of aircraft, navigation systems and procedures.[3]

Security: "Safeguarding International Civil Aviation Against Acts of Unlawful Interference" (Annex 17) deals with preventive security measures (with respect to access control, aircraft, passengers and baggage, landside and cyber threats) and the management of response to acts of unlawful interference (e.g. prevention, response and exchange of information and reporting).[4]

This Annex requires each Member State to establish its own civil aviation security program with such additional security measures as may be proposed by other appropriate bodies. It also seeks to co-ordinate the activities of those involved in security programs. It is recognised that airline operators themselves have a primary responsibility for protecting their passengers, assets and revenues, and therefore Member States must ensure that carriers develop and implement effective complementary security programs compatible with those of the airports out of which they operate.

2 See the minutes from the sixth meeting of the ICAO/Environment Protection's Committee on Aviation Environmental Protection in 2004 (CAEP/6) (DOC 9836).
3 This annex is available at www.ICAO.int.
4 See www.ICAO.int.

Annex 17 has been through ten editions in response to changing technologies and the appearance of new threats. A specific Aviation Security (AVSEC) Panel is responsible for its constant update. After the events of September 11 of 2001, a whole new set of measures was introduced.

Countries can extend the regulations further than what ICAO stipulates, while maintaining Annex 17 as the minimum standard. For example, the United States' Department of Homeland Security agency, Transportation Security Administration (TSA) has a whole set of specifications that go beyond Annex 17, applicable to all airports in US territory.

Facilitation (Annex 9) pertains to entry and departure of aircraft (e.g. documentation, inspection, procedures), persons, baggage and cargo, inadmissible persons and deportees, facilities and services for traffic at international airports, landing elsewhere than at international airports, facilitation provisions covering specific subjects (e.g. relief flights, marine pollution and safety emergency operations, health provisions, facilitation of the transport of persons with disabilities), and passenger data exchange systems (advance passenger information (API), electronic travel system (ETS), and passenger name records(PNR)).[5]

Annex 9 refers to facilities and services, in particularly with respect to traffic flow arrangements and facilities, handling of specific type of passengers, baggage and cargo, and facilities that have to be present at terminal buildings.

5.1.3 ICAO compliance

Regardless of the ownership of the airport, all aerodromes must obey the standards set forth by the Convention and its Annexes. ICAO recognise "standards" as necessary, and "recommendations" are desirable, and defines standards and recommendations as follows:

> A Standard is defined as any specification for physical characteristics, configuration, material, performance, personnel or procedure, the uniform application of which is recognized as necessary for the safety or regularity of international air navigation and to which Contracting States will conform in accordance with the Convention; in the event of impossibility of compliance, notification to the Council is compulsory under Article 38 of the Convention.
>
> A Recommended Practice is any specification for physical characteristics, configuration, material, performance, personnel or procedure, the uniform application of which is recognized as desirable in the interest of safety, regularity or efficiency of international air navigation, and to which Contracting States will endeavour to conform in accordance with the Convention. States are invited to inform the Council of non-compliance.[6]

---

5 Again, see the ICAO's website for this annex (ICAO, Annex 9, 2017).
6 From the ICAO's notes on "Making an ICAO Standard" (2011), available on their website.

72 *Air transport regulation*

Airports may not be able to comply with specific standard or procedure. Article 38 of the Chicago Convention[7] states that whenever a Member State finds it impracticable to comply with any particular standard or procedure, the Member State must give notification to the Council. Deviations from the standards are considered as exemptions and are to be published by the technical regulatory authority of the respective Member State. For example, Male International Airport's runway is insufficient to meet ICAO requirements. The Civil Aviation Authority of Maldives requested an exemption, on the condition that the runway strip be provided as per ICAO requirements by a specific date, according to the master plan. The exemption was granted by Member State and published in the Aeronautical Information Publication (AIP).

5.2 International Air Transport Association (IATA)

IATA is the trade association for the world's airlines, representing approximately 290 carriers and accounting for 82% of the global air traffic. The organisation advocates for the interests of airlines across the globe, challenging rules and charges they deem unreasonable while advising airports, governments and aeronautical entities on aviation policy and cost effectiveness.[8]

IATA offers a wide range of products and services, providing accreditation to travel agents and cargo agents, facilitating business interactions between agents and airlines and, most importantly, providing access to IATA's Billing and Settlement Plan (BSP) – an interface for invoicing and payment between the agent, airlines and transport providers. IATA also provides financial services, facilitating payment settlement systems and a clearing house to move funds between airports, airlines, travel agents, civil aviation authorities and air navigation service providers. Among them is the Enhancement and Financing Services (E&F) which is vital for collection of airport fees and charges. Other services that IATA provides include consulting services, safety and flight operations solutions, business intelligence and statistics and security solutions.

5.2.1 Services

IATA offers services in five areas:

1. *Consulting*: air service development and building new markets, terminal optimisation, traffic studies and transaction advisory services (including due diligence)
2. *Business intelligence*: market assessment and analysis tools, as well as the Aviation Charges Intelligence Center (formerly the airport, air traffic control and monitoring fuel charges) which covers over 50 aeronautical charge

---

7 Properly the Convention on International Civil Aviation, signed at Chicago, 7 December 1944.
8 International Air Transport Association (2019).

Air transport regulation   73

types from more than 670 airports, ATC charges for 200+ countries and jet fuel charges for 400+ airports
3. *Publications*: the following are IATA's main airport solutions publications:
   - *Airport Handling Manual* (AHM) with the latest industry-approved policies, standards and safety guidelines covering all facets of safe and efficient airport operations
   - *Ground Operations Manual* (IGOM) for all ground operations, providing standard operating procedures for frontline personnel
   - *Airport Development Reference Manual* (ADRM), long considered the gold standard for airports planning infrastructure, is published jointly with Airports Council International (ACI) and in close collaboration with world-renowned industry specialists
   - *Baggage Reference Manual* (BRM): to provide insight into the key features of baggage operations, the BRM presents potential issues that might occur along a generic baggage chain and ways to overcome them
   - *Unit Load Devices Regulations* (ULDR): technical/operational specifications and regulatory/airline requirements applicable to ULD operations
   - *Security Manual*: a tool for all airport security professionals, with updated security information and essential guidance on both current and emerging issues
4. IATA Work Groups
5. Training services.

Of them all, the *Airport Development Reference Manual* is of particular relevance for airport design and regulation.

5.2.2 Airport Development Reference Manual (ADRM)

The IATA Airport Development Reference Manual is the industry's most important guide for airlines, airports, government authorities, architects and engineering consultants who are either planning on constructing new or extending existing airport facilities. The ADRM has become the industry standard for airport planners worldwide as the primary source of best practice airport design guidance, particularly with respect to the level of service to be provided to passengers.

The ADRM covers the following areas:[9]

- Forecasting
- Planning (master planning, airside infrastructure, passenger terminal, cargo terminal, airport support elements, surface access systems, airport simulation, airport technology, jet fuel infrastructure, and operational readiness and airport transfer)

---

9 IATA Airport Development Reference Manual – 11th edition, March 2019b.

The ADRM is the world's reference for setting the required level of service of an airport. IATA's Level of Service (LoS) is commonly referenced as a planning guidance to define what is expected from a passenger terminal building. The LoS criteria takes into account objective parameters (such as terminal dimensions, facilities, available infrastructure, time delays) to determine an overall score for the airport terminal. It is also the most reliable approach to measure the offered level of service than passenger surveys. Utilising IATA's LoS indicators, two airports with the same infrastructure cannot have different scores (passengers' perceptions are not taken into consideration, only objective parameters).

The LoS framework considers the various terminal sub-systems and specifies the minimum service requirements at each, by way of space provision available to passengers as well as processing waiting time. These parameters can vary from one sub-system to the other: for example, the space requirement at passport control areas is different from pre-boarding areas. It is also often used for performance comparisons or as a benchmark that determines whether contractual obligations of airport owners, operators and/or third-party service providers are being met.

In recent years, IATA has adopted a new categorisation for LoS. In the updated ADRM, the LoS framework has been completely revised to better reflect both the dynamic nature of terminal operations and throughput as well as the intention of increasing infrastructure efficiency. The revised LoS Concept is based on four distinct categories:[10]

- Over-design
- Optimum
- Sub-optimum
- Under-provided

For LoS assessment purposes, the new framework now uses a two-dimensional LoS Matrix. The resulting LoS category is jointly dictated by two variables:

- Processing facilities: Space + Waiting time
- Holding facilities: Space + Occupancy

The general objective of IATA's new LoS philosophy is the provision of optimum passenger facilities, avoiding over- or under-provision of infrastructure. Terminal facilities that are designed according to LoS Optimum typically:

- Provide sufficient space to accommodate all necessary functions in a comfortable environment
- Provide stable passenger flows with acceptable waiting times

---

10 IATA, "Improved Level of Service Concept" (2017a); the information in this section is all drawn from this publication.

Table 5.1 IATA level of service assessment matrix

		SPACE		
		Over-Design	*Optimum*	*Sub-optimum*
MAXIMUM WAITING TIME	*Over-design*	OVER-DESIGN	Optimum	SUB-OPTIMUM Consider improvements
	Optimum	Optimum	OPTIMUM	SUB-OPTIMUM Consider improvements
	Sub-optimum	SUB-OPTIMUM Consider improvements	SUB-OPTIMUM Consider improvements	UNDER-PROVIDED Reconfigure

Source: "Improved Level of Service Concept" (IATA Consulting, 2017a)

- Deliver an overall good service (comfort level) to passengers while keeping capital expenditures and operating expenses at a reasonable level
- Balance economic terminal dimensions with passenger expectations.

The LoS has been modified to better reflect the current aviation market from a global perspective: Different regions, countries and markets require approaches to the airport environment to match their specific service needs. For this reason, the LoS parameters for "Optimum" are provided as a range of values (for space, waiting time and occupancy). This flexibility allows an airport to better tailor its service level to the market and region it serves.

The appropriate (or targeted) LoS value should be established through a proper consultation process involving all relevant airport stakeholders such as airport operator, airlines and other relevant service providers.

PPPs in airports shift the responsibility of service provision from the public sector to the private, necessitating a monitoring mechanism to ensure that a level of service is maintained by the private operator. Currently, most PPP contracts incorporate clauses dealing with future expansion requirements and planning of infrastructure in compliance with specific standards of service and minimum technical requirements. IATA's LoS have become the most common guidelines for airport planning and provision of quality of service.

5.2.3 IATA's enhancement and financing services (E&F)

Created in 1991, IATA's E&F Services aim to improve invoicing and collection of aeronautical charges for airports and air navigation service providers (ANSPs), and is currently the main third-party institution for collection of fees and charges worldwide. Fees and charges levied to passengers (PFC, AIF, security, etc.)

Table 5.2 IATA level of service guidelines

LoS Guidelines		SPACE GUIDELINES (sqm/PAX)			MAX. WAITING TIME GUIDELINES Economy Class (minutes)		
LoS parameters		Over-design	Optimum	Sub-optimum	Over-design	Optimum	Sub-optimum
Public departure hall		> 2.3	2.0–2.3	< 2.0	n/a		
Check-in	Self-service kiosk (Boarding Pass / Bag Tagging)	> 1.8	1.3–1.8	< 1.3	< 1	1–2	> 2
	Bag drop desk (queue width 1.4–1.6 m)	> 1.8	1.3–1.8	< 1.3	< 1	1–5	> 5
	Check-in desk (queue width 1.4–1.6 m)	> 1.8	1.3–1.8	< 1.3	< 10	10–20	> 20
Security Control (queue width: 1.2 m)		> 1.2	1.0–1.2	< 1.0	< 5	5–10	> 10

Source: "Improved Level of Service Concept" (IATA Consulting, 2017a)

through flight tickets are collected by IATA on behalf of the airports. In addition, fees and charges levied on airlines for the use of airport facilities (landing fees, use of counters, CUTE, etc.) are also collected through this facility. ANSPs, which charge airlines for services such as overflight (en route) approach, also engage IATA for the collection of funds from the airlines.

IATA E&F acts as an intermediary between airports and ANSPs, consolidating the operational information (which is provided by the airports), sending invoices to and receiving payments from the airlines, and finally forwarding these payments to the airports.

For airport services, the levying authority must sign one agreement with IATA and another with the airline community on the charges to be levied. IATA's invoicing and collection methods are not limited to IATA members. When an airline is not an IATA member, the IATA Settlement Systems for collection cannot be used, so instead IATA collects through direct payments (bank transfers).

IATA's E&F currently handles over 100 different charges on behalf of more than 120 airports and 30 FIRs (flight information regions) and invoices more than 4,000 different operators all around the world.

IATA, however, does not guarantee payment. IATA tries to make the process of invoicing and collection as efficient and automated as possible and therefore typically has collection rates of over 95%. The collection of fees and charges through the E&F facility provides a great deal of confidence to banks financing airport infrastructure.

5.3 Market access and air service agreements

The ICAO Convention sets out basic parameters for Member State negotiations on airline services.

5.3.1 Bilateral air service agreements

Countries fix their relations with respect to air transport by *Air Service Agreements* (ASAs) or *Bilateral ASAs* (BASAs). A BASA is an international trade agreement between two nations that sets the conditions under which air transport services between them will be established. They are negotiated by the institutions that set aviation policies in each country.

These agreements regulate several key aspects of air transport services:

- *Airline designation*: single or multiple designation of airlines; in the former, only one carrier per side may offer air services to the other country, while in the latter, several carriers can operate routes between the two countries
- *Capacity*: the level of capacity to be offered between the two countries, by defining the number of seats, or the number of flights, and may include a specification of the largest aircraft type allowed on a given frequency
- *Route schedule*: the exit points from one country and the corresponding entry points in the other country. The routes to be operated between the two countries are subject to the mutual grant of specific rights of operation or "freedoms" (called "freedoms of the air"). Bilateral agreements may involve a third country, specified in the route schedule as an *intermediate* point or as a point *beyond* the route destination. In specific cases, airlines may also be able to transport revenue traffic between those points, including the intermediate and beyond points[11]
- *Airfares*: either an indicative airfare for the different routes, a ceiling or a floor, or none. The current trend is not to intervene in commercial decisions of the carriers and to refrain from specifying airfare limitations. In some cases, airfares are considered accepted unless they are contested by the aviation authorities of both countries (called "double disapproval"). The most restrictive type of agreements may require both countries to approve a specific fare ("double approval").

Box 5.1 Freedoms of the Air

Traditionally, airlines need government approval of the various countries involved before it can fly in or out of a country, or even fly over another country without landing. Prior to World War II, this did not present too

---

11 This is known as a Fifth Freedom right.

many difficulties since the range of commercial planes was limited and air transport networks were in their infancy and nationally oriented. In 1944, an International Convention was held in Chicago to establish the framework for all future bilateral and multilateral agreements for the use of international air space. Five freedom rights were designed, but a multilateral agreement went only as far as the first two freedoms (right to overfly and right to make a technical stop). The first five freedoms are regularly exchanged between pairs of countries in Air Service Agreements. Freedoms are not automatically granted to an airline as a right; they are privileges that have to be negotiated and can be the object of political pressures. All other freedoms have to be negotiated by bilateral agreements, such as the 1946 agreement between the US and the UK, which permitted limited "fifth freedom" rights. The 1944 Convention has been extended since then, and there are currently nine different freedoms:

- *First Freedom*: the freedom to overfly a foreign country from a home country en route to another without landing. (for example, although without being able to land, US carriers have granted the right to overfly Cuba on the way to Latin America)
- *Second Freedom*: the freedom to stop in a foreign country for technical purposes only. A flight from a home country can land in another country for purposes other than carrying passengers, such as refuelling, maintenance or emergencies. The final destination is country
- *Third Freedom*: the freedom to carry traffic from a home country to another country for the purpose of commercial services
- *Fourth Freedom*: the freedom to pick up traffic from another country to a home country for purposes of commercial services.

The Third and Fourth Freedoms are the basis for direct commercial services, providing the rights to load and unload passengers, mail and freight in another country.

- *Fifth Freedom*: the freedom to carry traffic between two foreign countries on a flight that either originated in or is destined for the carrier's home country. It enables airlines to carry passengers from a home country to another intermediate country, and then fly on to a third country with the right to pick passengers in the intermediate country. Also referred to as "beyond right". This freedom is divided into two categories: *Intermediate Fifth Freedom* is the right to carry from the third country to the second country; *Beyond Fifth Freedom* is the right to carry from the second country to the third country (for example, Singapore Airlines operates flights from Frankfurt to New York JFK as a continuation from its flight from Singapore Changi)

- *Sixth Freedom*: the "unofficial" freedom to carry traffic between two foreign countries via the carrier's home country by combining the third and fourth freedoms. Not formally part of the original Chicago Convention, it refers to the right to carry passengers between two countries through an airport in the home country. With the hubbing function of most air transport networks, this freedom has become more common (most European and Gulf carriers operate a great deal of their traffic by using the sixth freedom)
- *Seventh Freedom*: the freedom to base aircraft in a foreign country for use on international services, establishing a de facto foreign hub. Covers the right to operate passenger services between two countries outside the home country (this is typical in cargo operations, for example FedEx basing operations in Paris to consolidate traffic from Europe and the Middle East)
- *Eighth Freedom*: the freedom to carry traffic between two domestic points in a foreign country on a flight that either originated in or is destined for the carrier's home country. Also referred to as "cabotage" privileges. It involves the right to move passengers on a route from a home country to a destination country that uses more than one stop along which passengers may be loaded and unloaded
- *Ninth Freedom*: the freedom to carry traffic between two domestic points in a foreign country. Also referred to as "full cabotage". It involves the right of a home country to move passengers within another country. In an extraordinary agreement, Chile and Uruguay have granted the ninth freedom reciprocally, under which in theory, an airline from Uruguay could be based in Santiago and operate flights domestically.

In addition, BASAs often include other commercial issues such as commercial arrangement between airlines -the possibility for carriers to share flight codes ("code sharing") or even specific joint ventures between airlines. BASAs may deal separately with charter and cargo flights. BASAs may also regulate the repatriation of funds, the provision of services at airports and the use of computer reservation systems (CRS).

Liberalisation

The level of definition of the four basic elements of an air service agreement specifies how restrictive or liberal an agreement can be. A restrictive agreement will designate one or a few airlines per country, while a very liberal one will leave open the number of airlines.

A restrictive agreement will specify a few individual points in each country, whether cities or even particular airports. A liberal agreement would define "any point in Country A, to any point in Country B" and vice-versa.

80 Air transport regulation

Table 5.3 Example of a BASA route schedule

Routes to be operated by the designated airlines of Country A

Origin	Intermediate points	Destinations	Points Beyond
Points in country A	Any points	Points in Country B	Any points

Routes to be operated by the designated airlines of Country B

Origin	Intermediate points	Destinations	Points Beyond
Points in country B	Any points	Points in Country A	Any points

Source: Authors

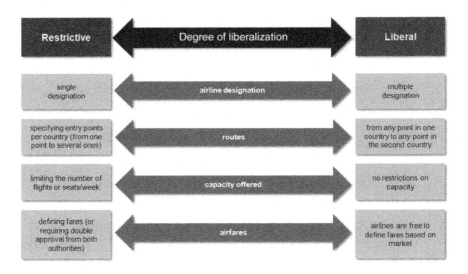

Figure 5.1 Degree of liberalisation of a bilateral air service agreement

Capacity. A limiting BASA will restrict the capacity to be offered, either by number of seats, or by number of flights and eventually combined with a maximum size of aircraft. A liberal BASA will not limit the capacity to be offered, leaving it completely to the market. Under some protectionist environments, one party may request effective reciprocity in terms of capacity, meaning that it would not allow the other party to offer more capacity than they themselves can offer.

Airfares. A restrictive agreement would either set an airfare (or could be a minimum, a maximum or a fare band), or require the airfare to be approved by the authorities of both countries (double approval). A liberal agreement will not restrict airfares. Another liberal approach would accept any airfare proposed by the market unless it is disapproved by both parties (double disapproval).

The development of air transport is directly related to the level of liberalisation of traffic rights between countries. Governments that have adopted a liberalised aviation policy have experienced sharp increases in traffic.

Box 5.2 Open Skies Agreements

Under Open Skies Agreements the four basic elements of any air service agreement are fully liberalised:

1. No specific designation of airlines
2. The route schedule doesn't have to be specified and any point in one country could be linked to any point in the other, allowing points before and beyond both countries
3. The capacity to be offered by each country is not limited
4. Airfares are subject to market forces and can only be limited by double disapproval of both countries under special conditions pertaining to competition issues.

These types of agreement, however, do not include the right of an airline of one country to perform commercial transportation services within the territory of the other country.

Since the early 1980s, the US Department of Transportation (DOT), under a mandate from the US Congress, has signed over a hundred such agreements with countries all over the world. Many countries around the world followed this policy by designing their own open skies agreements under their own terms, but maintaining the basic components of the US agreement.

5.3.2 *Multilateral air service agreements (MASAs)*

Multilateral ASAs are signed by three or more states, in which all signatory parties agree to provide each other with the same capacity, designation, airfare setting and route scheduling rights to operate air services. There are several MASAs in place at present, usually negotiated between countries located within the same geographical region.

The key features of the MASAs are:[12]

- Multiple airline designation
- An open route schedule, allowing airlines to fly from any point of their country, to any point in the other

---

12 These are drawn from the Government of New Zealand's MALIAT website (www.maliat.govt.nz/), retrieved on May 2019.

- Open capacity, with no restrictions on capacity
- Open traffic rights including seventh freedom cargo services, meaning that cargo airlines can fly out of any of the other countries and not as a continuation of a flight from the country of the airline
- Airline investment provisions which focus on effective control and principal place of business, but protect against flag of convenience carriers
- Third-country code-sharing.

Multilateral Agreement on the Liberalization of International Air Transportation (MALIAT)

The MALIAT was signed in Washington DC in 2001 by Brunei Darussalam, Chile, New Zealand, Singapore and the US to promote open skies air services.

ASEAN Multilateral Agreement on Air Services of the Association of Southeast Asian Nations

The Governments of Brunei Darussalam, Cambodia, Indonesia, Laos, Malaysia, Myanmar, Philippines, Singapore, Thailand and Vietnam, Member States of the Association of Southeast Asian Nations (ASEAN), signed a MASA in October of 2009.

The agreement establishes that:[13]

- Each country shall have the right to designate as many airlines as it wishes for the purpose of conducting international air services
- The air fares to be applied by the designated airline or airlines of a Contracting Party for air services shall be established at reasonable levels. The Contracting Parties agree to give particular attention to fares that may be objectionable because they appear unreasonably discriminatory, unduly high or restrictive because of the abuse of a dominant position, or artificially low because of direct or indirect governmental subsidy or support or other anti-competitive practices
- Substantial ownership and effective control of designated airlines are vested in the Contracting Party designating the airline, nationals of that Contracting Party, or both. The designated airline which is incorporated and has its principal place of business in the territory of the Contracting Party that designates the airline, is and remains substantially owned and effectively controlled by one or more ASEAN Member States and/or its nationals, and the Contracting Party designating the airline has and maintains effective regulatory control
- Each designated airline shall have a fair and equal opportunity to compete in providing the international air services governed by this Agreement

---

13 ASEAN Multilateral Agreement on Air Services, Manila, 20 May 2009.

- Liberated capacity for all intra-regional routes
- Liberated route schedules for third and fourth freedoms of the air
- In 2011, as part of the agreement, a full liberalisation of fifth freedom of the air between all capital cities of the region was set in place.

The agreement initially set a timeline to constitute a Single Aviation Market (SAM) with an open skies policy by 2015. However, the arrangement to form a SAM has been delayed by years of political negotiations between the Contracting Parties. According to ASEAN data, the total seat capacity of ASEAN airlines experienced double-digit growth in the four-year period of 2009–2013, and the share of low-cost carriers (LCC) in the region increased significantly from 13.2% in 2003 to 57% in 2014.[14]

The liberalisation of the aviation market has also provided increased competition, particularly as a result of the emergence of Low Cost Carriers (LCCs) in the region, which has reduced passenger air fares and facilitated greater choice. A total of 1,009 nonstop services were served in 2004 from ASEAN countries, which had increased to 1,683 nonstop services by 2014 – an increase of 674 routes (or 67%) in ten years.[15]

CARICOM Multilateral Air Service Agreement of the Caribbean Community

The MASA focuses on the exchange of route and traffic rights within the Caribbean Community and provides the required regulatory framework within which a CARICOM air carrier (an air carrier which is registered in a Member State, the majority of whose shares are owned by one or more Member States and/or their nationals) can provide air transport services between signatory nations.[16]

The main characteristics of the CARICOM MASA are:[17]

- The Agreement concerns the operation only of CARICOM air carriers. It allows all types of air services to be performed within the Community by those carriers designated by Member States
- More than one carrier can be designated by a country to exercise the traffic rights granted under the Agreement
- The traffic rights covered by the Agreement include the right to carry traffic between a country in which the carrier is registered and another country; and on a reciprocal basis, the right to carry traffic between another Contracting State and beyond

---

14 This data is from ASEAN's "Building the ASEAN Community, ASEAN Single Aviation Market" (2015).
15 Data from the Institute for Democracy and Foreign Affairs, "Economic Benefits of ASEAN Single Aviation Market" (2018).
16 See the ICAO's "Liberalization of Air Transport Services within the Caribbean Community" (2003).
17 *Ibid.*

- There is no obligation for a Member State to grant cabotage traffic rights to the carrier of another party, neither is there a prohibition to grant such rights
- Airfares in respect of scheduled air services are required to be submitted for approval and are deemed to be approved if neither of the two States concerned expresses its disapproval of the fares within a specified period
- The Agreement provides for a fair and equal opportunity for all CARICOM air carriers to compete in the air transportation covered by the Agreement
- There is no provision for the direct control of capacity on any route. However, the undesirable practice of "dumping" excessive capacity in order to force a competitor out of business can be addressed on the basis of the commitment of Member States to eliminate unfair competitive practices and to have as their primary objective, the provision (at a reasonable load factor) of capacity adequate to meet the current and reasonably anticipated traffic requirements.

Decision on Integration of Air Transport of the Andean Community (CAN)

Signed in 1991 by Bolivia, Colombia, Ecuador and Peru, this MASA established the conditions for the establishment of intra-regional international air services. The agreement established the following main dispositions:[18]

- The Member States grant each other the free exercise of the third, fourth and fifth air freedoms for regular passenger, cargo and mail flights within the Subregion
- The Member States accept the principle of multiple designation for the provision of regular passenger, cargo and mail services
- The corresponding Competent National Agencies shall automatically grant authorisations to provide unscheduled passenger, cargo and mail air transport services within the Subregion by the national carriers of the Member States
- The Andean Committee of Aeronautical Authorities, created by Resolution V.104 of the V Meeting of Ministers of Transportation, Communications and Public Works of the Member States, shall be responsible for ensuring integral compliance with and application of this agreement
- The Andean Committee of Aeronautical Authorities shall be comprised of the authority responsible for civil aviation in each Member States and his alternate, who shall act as the Titular and Alternative Representatives of that country, respectively, and shall be accredited before the Board.

---

18 See the Organization of American States, "Decision 297: Integration of Air Transport in the Andean Subregion" (1991).

The EU Internal Market and the European Common Aviation Area (ECAA)

Following the Maastricht Treaty that established the European Union, a common Internal Market for aviation was created in 1992. From a highly regulated industry, dominated by national flag carriers and state-owned airports, the internal market was developed into a common airspace without commercial restrictions for airlines flying within the EU.

The regulatory framework for air services ("the Air Service Regulation": Regulation 1008/2008) provided the economic framework for air transport in the European Community that provides the economic framework for air transport in the European Community setting out the rules:[19]

- The grant and oversight of operating licenses of Community air carriers
- Market access
- Aircraft registration and leasing
- Public service obligations
- Traffic distribution between airports; and
- Pricing.

EU air carriers can operate on any route with EU states, without restrictions of capacity or price control. Any airline registered in the EU is automatically granted all nine freedoms of the air between Member States.

Since this MASA, the integrated common airspace traffic has proliferated significantly. The introduction of new routes by LCCs and legacy carriers has dramatically reduced airfares while levels of quality of service and safety have increased.

European Common Aviation Area (ECAA) Agreement

Signed on 9 June 2006 and entered into force on 1 December 2017, the ECAA Agreement is a MASA that integrates all the countries in the Western Balkan region into the European aviation area. The agreement was signed between the European Union, Norway and Iceland, and the partners from South-Eastern Europe: Albania, Bosnia and Herzegovina, North Macedonia, Montenegro, Serbia, Kosovo.[20]

The aim of this aviation agreement is the creation of a European Common Aviation Area (ECAA), integrating the EU's neighbours in South-East Europe into the EU's internal aviation market which consists of EU Member States as well as Norway and Iceland. The ECAA agreement, by creating a single aviation market, should deliver substantial economic benefits for air travellers and the

---

19 See the *Official Journal of the European Union*, Regulation (EC) No 1008/2008 Of the European Parliament and of the Council of 24 September 2008, on common rules for the operation of air services in the Community (Recast).
20 See the *Official Journal of the European Union*, 9 June 2006 (2006/682/EC).

86 Air transport regulation

aviation industry, covering 36 countries and more than 500 million people. At the same time, the agreement applies the same high standards in terms of safety and security across Europe, through the uniform application of rules.[21]

5.4 Slot allocation

Airports that face capacity constraints that cannot satisfy traffic demand are required to set rules for the allocation of specific timings for landings and take-offs. The formulation of these rules is known as "slot coordination". IATA defines a *slot* as "a permission given by a coordinator for a planned operation to use the full range of airport infrastructure necessary to arrive or depart at a Level 3 airport on a specific date and time".[22] The reference to "Level 3 airports" means those airports

> where capacity providers have not developed sufficient infrastructure, or where governments have imposed conditions that make it impossible to meet demand. A coordinator is appointed to allocate slots to airlines and other aircraft operators using or planning to use the airport as a means of managing the declared capacity.

According to the International Civil Aviation Organization (ICAO), a slot represents "the approval that an airline needs to access the full range of airport infrastructure necessary for an aircraft to arrive at or depart from an airport on a specific date and time."[23]

Capacity constraint may occur only at certain periods of the day or on certain days of the week, or even during specific seasons. The allocation of a slot at a particular airport has implications in terms of market access and operation of international air services.

The situation varies widely across regions. According to the International Air Transport Association (IATA), the total number of capacity constrained airports that have been labelled as "fully coordinated" or subject to slot allocation under the IATA Schedule Coordination System continues to increase. As of May 2019, over 200 airports in the world were Level 3.

In addition, over 100 airports are considered Level 2, where "there is potential for congestion during some periods of the day, week or season, which can be resolved by schedule adjustments mutually agreed between the airlines and facilitator".[24]

---

21 European Commission – Mobility and Transport – 1 June 2019.
22 IATA, "Worldwide Slot Guidance" (2017b).
23 ICAO A39-WP/3401 EC/33 30/8/16.
24 In contrast, a Level 1 airport is one where the capacity of the airport infrastructure is generally adequate to meet the demands of airport users at all times (IATA, "Worldwide Slot Guidance", 2017b).

IATA defines Worldwide Slot Guidelines (WSG) to provide the global air transport community with a single set of standards for the management of slots at coordinated airports and planned operations at facilitated airports. The WSG is the industry standard recognised by many regulatory authorities for the management and allocation of airport capacity.

Restrictions on access to a particular airport because of lack of availability of slots may supersede the capacity granted under ASAs. Foreign airlines that may have given rights to fly into London, for example, may not be able to fly into London Heathrow because of the lack of available slots. During renegotiations of the ASA between the UK and South Africa, the Department of Transport of South Africa was reluctant to grant additional frequencies to British Airways into Johannesburg O.R. Tambo International Airport. The UK DoT's argument was that the South Africa flag carrier, South African Airways, was not going to be able to implement its reciprocal rights because of the scarcity of slots at Heathrow airport.

In fact, the ICAO provides specific guidance to prevent the manipulation of the procedures for slot allocation, stating that "allocation should be fair, non-discriminatory and transparent, and should take into account the interests of all stakeholders while it should also be globally compatible, aimed at maximizing effective use of airport capacity, simple, practicable and economically sustainable"[25].

Slots may have an intrinsic value, constituting an asset for airlines, depending on the specific rules at the airports. In the UK, the slot allocation process is carried out according to the IATA guidelines and EC Regulation (95/93)[26] for Heathrow, Gatwick, Manchester, Stansted and London City Airports.[27] The process of slot allocation takes place at airport capacity meetings at a Slot Coordination Committee. It involves the consideration of capacity parameters (runway, terminals, number of stands, etc.) between the airport slot managing body, the air navigation service provider (NATS) and the airlines. The slots are allocated according to specific criteria, for which historic precedence is the primarily criterion. Second, some slots have historic rights but have changed in terms of type of aircraft or timing. New entrants and new incumbents are also considered, according to the different status of new airlines or new services. Other considerations such as the size of the market, competition issues and network scheduling limitations are also considered.[28]

At Sydney Airport, slot allocation is ruled by the Sydney Airport Demand Management Act of 1997. A Slot Manager is appointed and receives recommendations from a Compliance Committee, with members appointed by the Minister. The main criteria used to allocate slots include prioritising historic

---

25 "Worldwide Air Transport Conference (Atconf): Slot Allocation" (ICAO, 2012a).
26 *Official Journal of the European Union*, Council Regulation (EC) No 95/93, 18 January 1993 on common rules for the allocation of slots at Community airports.
27 "UK Slot Coordination Process and Criteria" (Airport Coordination Limited, 2016).
28 *Ibid.*

slots over new requests, promoting regional routes by ensuring services are either not interrupted or introduced as new, prioritising larger over smaller aircraft, and prioritising year-round services rather than seasonal.[29]

The possibility to swap or transfer slots between air carriers means that a slot has an inherent asset value. For example, the EC regulation allows slots to be exchanged or transferred between airlines in cases of partial or total takeover, or transfer to a different route or traffic mode. Such exchanges or transfers must be transparent and subject to confirmation from the coordinator.

Exercising this right, British Airways' parent company, International Airlines Group (IAG), paid an estimated €100m to €150m to Lufthansa for six pairs of slots at Heathrow, belonging to its subsidiary British Midland International (BMI) in 2011. Continental Airlines, in 2008, paid USD 209m for four pairs of slots at Heathrow to be used for services to Houston and Newark airports.[30]

However, regulations on slot allocation and trading are subject to antitrust rules. In the case of BMI's takeover by the British-Spanish group of companies, the European Commission forced IAG to give up 14 of the 56 pairs of slots acquired through the takeover of BMI. The acquisition would have increased IAG's share of Heathrow slots from 45% to 51%.

5.5 Economic regulation of airports

Fees, charges and levels of service at airports managed by the public sector are a matter of national aviation policy. Governments need to safeguard the interests of users, to ensure that the users are not charged excessively and that services meet expectations, but also to ensure profitability of airports, to fund maintenance and investment. To avoid political or commercial capture of fees and charges, most countries establish an independent regulator. Economic regulation in general comprises regulations, schemes and procedures that define the economic structure of an airport, the fee/charge structure and the form in which it is imposed. This section discusses different aspects of economic regulation.

5.5.1 Non-price control regulatory regimes

There are four non price-control economic regulation regimes that are commonly found at most airports:

1. Market place
2. Ministerial permission
3. Voluntary agreements
4. Reserve powers.

---

29 Sydney Airport Demand Management Act of 1997.
30 "BA Buys BMI Heathrow Slots", *Financial Times*, September 23, 2011.

The *market place* determines charges levied by airports by means of negotiation between the airport and its users, namely the airlines. This relationship is a symbiotic one rather than a monopoly. This kind of regulation can be seen in many unconstrained regional airports.

Under *ministerial permission,* airports seek annual permission through a consultative process with their respective Ministries to alter airport charges. This type of economic regulation is often characterised by intensive lobbying mechanisms either by airports themselves or by national airline companies. The process is highly politicised and is often considered a serious obstacle for investors. This type of economic regulation is typical for airports run by state-owned entities as the role of the state is decisive.

Voluntary agreement involves a collaborative arrangement between the airport and its users, commonly for rather short terms such as three to five years. These agreements involve comprehensive analysis of levels of investments and associated standards for services. The dialogue between the airport and users is based on the justification of the needed investments to support the increase in charges. Copenhagen and Frankfurt use this type of agreement.[31]

Reserve powers uses legislation to provide a strict price control mechanism. This implies monitoring of prices, accounts and service standards by the existing regulatory body/agency. Examples of this type of regulation can be found in Australia and the UK.

5.5.2 Price monitoring regulation

Under price monitoring, airports are not subject to direct price regulation. Instead, they are subject to price monitoring, typically for a period of five years, after which there is a review of performance. If the airport's performance has not been satisfactory it may be re-regulated. Price monitoring regulation provides little or no incentive to the airport to consult with the airlines. Instead of reaching higher efficiency levels, the goal of the airport is to avoid excessive price increases in order to fly under the radar of the regulator.

5.5.3 Cost-based regimes

Cost-based regulation sets prices that reflect the costs of the regulated firm. Rate of return regulation requires the regulator to set prices in order to deliver a "reasonable" rate of return on invested capital. The objective is to obtain an ongoing service without allowing for excessive profits.

There is a risk that, if for some reason the rate of return is set too high, the airport would have a clear incentive to increase its capital expenditure

---

31 As an example, in the past Fraport has reached an agreement with airlines operating at Frankfurt under which they accepted to pay more, helping to finance a multi-billion euro expansion (Reuters, February 20, 2010).

inefficiently. This would result in an increase in prices, for services that the users do not require.

The cost-based regulation regimes have complex information requirements. The regulator must set fair charges based on information provided by the company regarding operating expenses, capital expenditure, volume projections and asset base. It must also assess whether the charges are efficient and appropriate.

Rate of Return regulation involves a mechanism whereby prices are adjusted to assure a target rate of return, based on the level of throughput. The construction and operation of Terminal 2B at Budapest-Ferihegy International Airport, for example, was regulated by this method. This method involves periodic price revisions in order to compare with the target rate of return.

The main issue with cost-based regulation is that the regulated firm will not benefit from cost-reducing efforts as any savings will also be passed through to customers. Prices are not linked to opportunity costs, thus it may not be appropriate to use this type of regulation for cases where there is congestion and new facilities need to be built.

5.5.4 Incentive-based regulation: Price cap regulation

Incentive-based regulation was introduced to overcome problems associated with the rate of return regulation and to provide incentives for firms to act in ways to improve economic welfare.

This regime is based on the use of price control by the implementation of price caps as the main incentive mechanisms used by regulators. Price caps specify the maximum price for a good or service over a certain period. Typically, price increases are constrained to a level determined by an index that reflects the exogenous increase in input costs – commonly the rate of inflation minus an X factor that accounts for expected productivity improvements. Any cost savings achieved beyond the index minus X accrue to the firm within a given regulated period.

For example, assume that the US CPI has increased by 3%, and that the X factor is set at 1%. This means that the airport would be allowed to increase regulated charges by only 2% (3%–1%). This means that in real terms, the charges would be decreasing, as may not accompany the inflation. The decrease in charges is a way of transferring the progressive economies of scale and learning curve gained by the operator, onto the users.

Besides the inherent advantages in terms of incentives and economic efficiency, price cap regulation provides the regulator with more degrees of freedom when compared to other regimes, because of the range of factors that can go into the X.

Price caps, when compared with cost-based regulation, provide more incentives for productive efficiency. They attempt to provide regulated firms with incentives for efficient supply and price structures, while at the same time encouraging firms to implement efficient price levels.

As mentioned above, cost-based regulation has the drawback of giving the firm an incentive to exaggerate both operating and capital costs. If the return is calculated on its asset base it will try to expand it, without evaluating the efficiency of such investments.

In terms of risk sharing, under a cost-based regulation, the risk of any investment is borne by the customers (or even the regulator), since the firm is allowed to pass through excessive costs to prices. Under the price cap regime instead the risk of investment is shared between the regulated firm and its customers, because a rate of return is not specifically guaranteed. Thus, the incentive of overcapitalisation is absent under price cap regulation.

But even though price cap regulation does not produce over-investment incentives, it is important to establish an effective consultation process with the users, in order to reach consensus about new investments in a spirit of openness, understanding, and collaboration from both sides.

5.6 Single till versus dual till

When setting aeronautical fees, government must decide whether non-aeronautical revenues should offset these fees. Two broad options are:

- *Single till*: aeronautical charges should be calculated considering profits earned from non-aeronautical activities as offsetting costs. This model forces cross-subsidisation of aeronautical activities from any above-normal non-aeronautical profits, and provides incentives for cost minimisation in the provision of non-aeronautical and aeronautical activities (during the period of the price cap)
- *Dual till*: under this option, aeronautical fees should not consider proceeds from non-aeronautical activities.

5.6.1 Effect on prices

Under the single till option airport costs are shared between aeronautical charges and commercial revenues, allowing the commercial activities to cross-subsidise the aeronautical ones. The dual till option implies that fees and charges for aeronautical services will have to cover the operational costs of the airport, including the capital expenditure. This usually has an impact on the level of fees and charges and generally increases user prices. These increases can be significant for passengers, particularly those passenger segments where elasticity of demand may reduce traffic (e.g. low cost carriers).

As airlines bring in customers and consequently make commercial operations possible, it is reasonable that they should also share in the benefits. Under the single till option, the airlines will indirectly receive a share of the airport's retailing profits in the form of lower aeronautical charges. This is consistent with the principle of treating the airlines as partners in the airport business, as opposed to the dual till option, where airlines are not treated as full partners.

Table 5.4 Single till versus dual till regulation

Single Till	Dual Till
Non-aeronautical revenues should be used to offset aeronautical charges	Non-aeronautical revenues should not be considered in the calculation of aeronautical charges
Supported mainly by airlines (including IATA) – also by many airport regulators	Supported by the airports (including ACI)
• Airlines should also benefit from non-aeronautical revenues since they bring the market to the airport • When airport congestion is not a major problem, single-till price-cap regulation dominates dual-till price-cap regulation with respect to overall social welfare * • No need for cost and asset allocation between aeronautical and non-aeronautical (which could be arbitrary under dual till – and difficult to regulate)	• Performs better than single-till regulation when there is significant airport congestion • Provides more incentives for investment when airports are performing in a competitive environment • Motivates airports to invest in aeronautical facilities (airside and pax processing areas at landside)

Source: authors; *Hangjun (Gavin) Yang, Anming Zhang, Sauder School of Business, UBC, 2011
Note: IATA: International Air Transport Association, ACI: Airports Council International

Cost allocation between aeronautical and non-aeronautical activities

It is generally conceptually difficult to separate commercial activities from aeronautical activities. The throughput of passengers at an airport is almost entirely dependent on aeronautical activities and so commercial activities based within the terminal are dependent on aeronautical ones (but not vice-versa). Commercial revenues cannot be generated without aeronautical facilities.

In the same fashion, it is difficult in practice to allocate investments and operating costs between aeronautical and commercial activities. This allocation could harm relations between the airport and its users. Under a dual till option the airport operator will have a strong incentive to assign all common costs to aeronautical activities, thus further increasing aeronautical charges.

5.6.2 Efficient use of services and congestion prices

It is not clear that the dual till would have a beneficial effect on efficiency in the utilisation of aeronautical facilities. It is argued that since single till prices reflect cost-sharing between the various commercial and aeronautical services, the demand for aeronautical services is greater than it would be if aeronautical services were charged on a stand-alone basis. Some scholars maintain that the single till might be perceived as inefficient, especially in congested airports.

The cross-subsidisation of aeronautical services reduces fees and charges to users, and increases demand for services for which there is a greater elasticity of demand,[32] for example the use of spaces inside the terminal building (offices, counters, VIP lounges, etc.), in other landside areas (car parking, hotels, gas stations, etc.) or spaces in the airside (aircraft maintenance, catering, cargo handling, etc.). Price reductions provide great incentives to users to increase their on-site presence.

In a competitive market, with no or few barriers to entry, congestion pricing (peak hour surcharges to try and change user behaviour) should lead to more efficient use of existing resources for the provision of aeronautical services and a signal for resource expansion. However, when demand elasticity is very low congestion pricing is likely to lead to:

- High rents being earned by the infrastructure provider
- Continuing use of scarce resources by users that cannot relocate services to another airport
- No incentive for the existing provider to expand services (or even a perverse incentive to maintain congestion).

There are significant barriers to entry in the provision of airport services, given the planning requirements for building new airports or extending existing ones. In principle, allowing an airport to charge congestion prices would allow airports to extract rents via higher user fees, most of which would be passed through to passengers through higher fares and some of which would be absorbed by airlines. Since price elasticity of demand is low, it is highly probable that passengers will bear the higher costs and as a result, the efficiency objective of congestion prices will not be achieved. Even if airlines were forced to bear the higher costs, the additional efficiency will be limited. The kind of constraints foreign airlines usually face include commitments from bilateral agreements, geopolitical motivations, corporate strategy implications, market appeal or the desire to maintain a presence. As such, the degree of efficiency savings that might be achieved is likely to be small, while a potentially large gain in rents to the airports could accrue.

5.6.3 Effect on investments

The dual till option could risk distorting investment incentives: commercial activities might expand at the expense of aeronautical, which may not attract sufficient funding or attention. In fact, there is no evidence that the single till option has led to under-investment in aeronautical assets. Moreover, the airport operator may benefit from having less space allocated to aeronautical needs because passengers would have to spend more time in the retail areas.

---

32 Services for which, under lower prices, demand would increase more than proportionally.

It is argued that separating the tills will strengthen investment in two ways:

- Pricing "correctly" should identify scarce resources and send appropriate signals for investment in new resources
- The allowed rate of return will cover aeronautical services rather than combined commercial and aeronautical services and thus should provide clearer incentives for investment.

One argument usually used to defend the dual till system is that by removing commercial activities from the aeronautical till, a potential barrier to efficient investment is removed. There is however no evidence that investment in commercial activities has been unduly hampered under the single till option. Under the single till option, airport operators are permitted to profit from any commercial revenues that exceed the assumptions made when setting the price cap. This is intended to encourage the airport to develop commercial potential, but not in ways that conflict with the main purpose of the airport or the interests of its users, as under the dual till option.

In addition, introducing the dual till option on the grounds of regulatory consistency is not compatible with ICAO guidelines that state that each airport should be considered and managed on its merits. Users do not want to pay the same no matter the airport facilities provided.

The Revenue Yield versus Charges Basket approaches

The two basic approaches to regulating airport fees are the *charges basket* and the *revenue yield*.

The *charges basket approach* uses a weighted sum of prices based on revenue shares in the previous period. The regulated airport is restricted to increasing the weighted sum of prices by no more than the CPI-X factor each period. Within the basket, the airport is able to increase or decrease individual prices, as long as they do not violate the weighted average sum price cap.

The *revenue yield approach* sets a price cap on a maximum allowable revenue yield per passenger or any other measurement unit. Under a single-till regime, the maximum allowable revenue yield per passenger is calculated by dividing total allowable revenue for aeronautical services by the anticipated number of passengers. The airport can set the prices for aeronautical services so as not to exceed the maximum average revenue yield per passenger. As with the charges basket, prices for individual airport services can be set freely by the airport as long as the average revenue constraint is not exceeded. Prior to doing so, however, the regulator will also want to ensure that ICAO principles are adhered to and that the airlines are consulted in this matter.

The charges basket method does not consider the possibility of new services, or dramatic changes in the composition of the airport product mix. These changes are typical of immature markets where there is still great potential for commercial development.

The *revenue yield* is a more transparent and easier to control mechanism, since it reduces the regulation to the total collected per unit of throughput and often uses the WLU mechanism (Work Load Unit, of one passenger or 100 kilograms of cargo). This mechanism translates to the airport operator the decision that has been made in terms of the distribution of fees and charges between the passengers and the airlines. The airport operator is controlled by its revenue per unit of throughput, but retains the discretion of the mix of fees and charges that it chooses to implement between all the regulated revenues.

The revenue yield approach tends to be more flexible because there are no restrictions on the charges that can be levied as long as the airport adheres to the average revenue constraint. Introducing new charges under the revenue yield is a straightforward process, whereas under the charges basket approach new weights should be determined when new charges need to be introduced.

The charges basket approach provides better incentives to move to efficient pricing methods. Under the revenue yield approach, when there is unused capacity, the airport may set prices below incremental costs for some services and recover them on other services.

In the UK the regulation of airports is via the revenue yield approach. The maximum allowable revenue yield per passenger is calculated by dividing the total allowable revenue by the anticipated number of passengers. This approach was used for Stansted and Manchester as well as for Heathrow and Gatwick.

The charges basket approach involves discretional weighting of charge increases, which can lead to differences in opinion in the criteria considered by each side for the weighting of charges. A revenue yield, instead, would leave freedom to the operator to distribute charges (and increases) between the regulated revenues. For example, the charges basket approach has been in practice in all three groups of airports privatised in Mexico (ASUR, GAP and OMA). The preferred approach remains however the revenue yield.

5.7 The UK example of regulation

The UK CAA adopts a price cap regulation when revising, for example, the fees and charges of London Heathrow and London Gatwick. This regulatory mechanism, also known as *RPI-X* limits the amount that can be levied by way of airport charges on a per passenger basis over the price control period.

This RPI-X mechanism is commonly employed for regulating public utilities and services. The X value is calculated by use of the *building blocks* approach. By this mechanism, the revenue requirement is calculated as the cost of capital or the return on the Revenue Asset Base (RAB) and the projected level of operating expenditures (opex). The RAB is the accumulated capital expenditures (historically purchased assets plus subsequent investments), adjusted for depreciation.

Under the Single Till approach, projected revenues from non-regulated charges are deducted from the total revenue requirement. It assumes that non-regulated

Table 5.5 Examples of single till, dual till and hybrid till regulation

Airport / Operator	Type of regulation	Type of till
London Heathrow	Price cap	Single till
Oslo	Cost based	Single till
Dublin	Price cap	Single till
Jo'burg O.R. Tambo (ACSA)	Price cap	Single till
Bogotá	Price cap	Single till
Lima (LAP)	Price cap	Dual till
Sydney Kingsford Smith	Price monitoring	Dual till
Vienna	Price cap	Dual till
Copenhagen	Airline contracts (price cap)	Dual till
Cancun (ASUR)	Price cap	Dual till
Frankfurt	Airline contracts (cost based)	Dual till
Hamburg	Price cap	Dual till
Athens Venizelos	Price cap	Dual till
Amsterdam Schiphol	Cost based	Dual till
Budapest Ferihegy	Price cap	Dual till
Spain (AENA)	Cost based	Dual till
New Zealand	Price monitoring	Dual till
Rome Fiumicino	Airports set own charges	Dual till
Brussels	Airline contracts (cost based)	Hybrid
Guarulhos (GRU)	Price cap	Hybrid
Paris (ADP)	Price cap	Hybrid
Stockholm Arlanda (Swedavia)	Cost based	Hybrid
Lisbon Portela (ANA/Vinci)	Price cap	Hybrid
Quito (Quiport)	Price cap	Hybrid
New Delhi (GMR)	Price cap	Hybrid

Source: compiled from airport regulatory bodies and airports financial statements

revenues (typically non-aeronautical or commercial revenues) contribute to the financing of the investment and the operations, by cross-subsidising the aeronautical activities. What is left after deducting these non-regulated revenues has to be recovered through regulated airport fees and charges.

The CAA considers this form of regulation appropriate given Heathrow and Gatwick airports' degree of market power. The RAB is a well-known model for regulation of organisations which have Substantial Market Power (SMP). For example, it is used in the UK in regulated sectors such as energy, water, rail, and wholesale telecommunications.[33]

The RAB balances the needs of users today and users in the future by ensuring that airport prices are no more than the minimum needed to remunerate an efficient airport operator, while providing a fair return on investment.[34]

---

33 "Economic Regulation at Heathrow from April 2014: Final proposals" (UK Civil Aviation Authority, 2014).
34 *Ibid.*

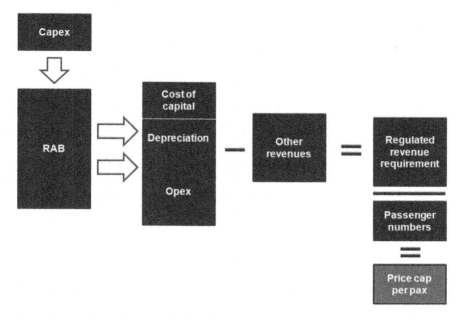

Figure 5.2 Building blocks to calculate price cap at UK regulated airports under single till regulation
Source: UK Civil Aviation Authority

5.8 Creation of a PPP monitoring body for economic regulation

Economic regulation should be vested in a body independent from political control. The body should be funded by the government but should be guided by statutory duties.

The regulatory body should safeguard the interests of the users (airlines, the traveling public and cargo owners), approving infrastructure development and master plans and ensuring the quality of services provided by airports. It should therefore be independent from airports and from airlines.

The provisions of ICAO's policies on charges, including the key charging principles of non-discrimination, cost-relatedness, transparency, and consultation with users, should be incorporated into the economic regulatory framework.[35]

The greatest risk when appointing a regulatory body is regulatory capture, where the regulatory body is influenced by political or business groups and acts in their interest rather than protecting the public interest.

---

35 "Policies on Charges for Airports and Air Navigation Service" (ICAO, 2012b).

When the number of those regulated is small, the relationship between the regulator and the regulated poses a challenge to the institutional strength of the country. It is harder for a regulator to maintain the necessary arm's length with the regulated as time goes by. One way of overcoming this challenge is by creating a multi-sector body that is in charge of regulating different sectors. The body will need to have specific specialised teams per each sector, but at board level is empowered by the responsibility to monitor various PPP contracts. The risk of being influenced by a particular regulated operator is significantly lower.

A good example of such a case is the Peruvian regulatory body created in 1998. OSITRAN is the regulatory agency for supervision of investments in transport infrastructure, is responsible for overseeing various sectors: airports, roads and motorways, railways (including metro), ports and waterways. It is a decentralised public body and is attached to Peru's Presidency of the Council of Ministers, having administrative, functional, technical, economic and financial autonomy.[36]

Its general objective is to regulate, regulate and supervise – within the scope of its competence – compliance with PPP concession contracts, protecting the interests of the State, investors and infrastructure users.

The declared objectives of OSITRAN[37] are to:

- Ensure the free access of end users to adequate infrastructure
- Guarantee quality and continuity in services of transport infrastructure
- Supervise the fulfilment of signed concession contracts
- Facilitate the development, modernisation and efficient use of transport infrastructure
- Promote competition in the supply of services
- Opine on charges, access to infrastructure and quality of service for new concession contracts.

As the economic regulator of airport concessions, OSITRAN oversees and sets aeronautical fees and charges and verifies the fulfilment of concession contracts, ensuring the achievement of the commitments assumed by the project company in terms of capital expenditure and level of service. In addition, it observes and imposes sanctions and corrective measures for breach of contract.

OSITRAN also resolves administrative disputes and claims that may arise between concessionaires or between these and their users.

In the UK, a specific regulatory unit in the Civil Aviation Authority (CAA) regulates the economics of airport PPPs and the provision of air navigation services. The unit operates at arm's length from the policy making and technical regulation teams of the CAA.

---

36 Lincoln Flor, Senior Transport Economist, World Bank (Former Manager for Regulation of OSITRAN).
37 From "Rol y funciones del Organismo Regulador", Patricia Benavente (OSITRAN, 2011).

Market power conditions are not met, therefore economic regulation is required to ensure protection of users. The CAA regulates the London airports of Heathrow and Gatwick. According to the CAA, the main reasons for regulating these airports are:[38]

- To prevent excessive pricing (above a competitive level)
- To prevent under- and inefficient investment
- To prevent poor efficiency
- To prevent poor service quality
- To prevent unfair trading conditions.

No airport PPP can function without an economic regulatory body responsible for monitoring the contract compliance. It is common practice that this function is vested in a body free from political interference and financial independence. Even when the body resides within a ministry, it is best custom to place it under a specialised professional unit.

All airports in Japan that were transferred under some sort of PPP are economically regulated by the Japan Ministry of Land, Infrastructure, Transport and Tourism (MLIT). In Chile, the Concessions Unit of the Ministry of Public Works (MOP) is responsible for overseeing the airports concession contracts and setting the level of charges.

Airports concessions in India are regulated economically by the Airports Economic Regulatory Authority (AERA), a statutory body constituted under the Airports Economic Regulatory Authority of India Act. In Australia, the economic regulation of privatised airports is carried out by the Department of Infrastructure, Transport, Cities and Regional Development. In Mexico, airports concessions are economically regulated by the Secretary of Transports and Communications (STC).

An effective economic regulatory body should employ minimal human resources – only economists, legal advisors and market analysts to monitor contractual compliance of a PPP. The body should rely on external consultative advice from other entities such as the antitrust regulatory agency and specialised commissions (like Australia's Productivity Commission). With respect to the technical aspects of airport regulation, the regulatory body will rely on advice from the technical regulatory body, typically the civil aviation authority.

---

38 "In January 2014, the CAA determined that the operators of Heathrow and Gatwick airports met the market power test set out in the Civil Aviation Act 2012, and therefore required a licence to recover charges for their services. These licences include conditions relating to prices, service quality and operational resilience, amongst others. Out of 60 airports in the UK, only Heathrow and Gatwick are subject to economic regulation." From "Best Practice in Economic Regulation: Lessons from the UK" (ICAO, 2014).

6 Airport PPPs around the world

> I'd never join a club that would allow a person like me to become a member.
> Groucho Marx (1890–1977), an American comedian and actor,
> from a telegram to the Friars Club of Beverly Hills, as recounted
> in his autobiography *Groucho and Me* (1959), p. 321

This chapter describes how PPP models are used to support airports, the types of PPP commonly used, and the approach to airport PPP used in countries and regions around the world.

6.1 Why airport PPP

Governments often ask "Why consider PPP? What are the merits of a PPP? Have they been successful?". There are basically three motivations for governments to consider PPP for an airport:

1. Policy drive to remove government from activities that are "private" in nature, allocating financially viable activities for privatisation
2. Mobilising private capital
3. Improving management.

Box 6.1 The case of the Government of Australia's reasoning for adopting a private management model

The Australian government decided to transfer most of their airports to private entities. (The operation of Australia's 22 federal airports was privatised between 1997 and 2003 by selling long-term leases over the airport sites to private sector operators.) Through the Airports Act of 1996, the Government of Australia set forth the way the private sector would be involved in the operations of airports, defining the following objectives:

(a) To promote the sound development of civil aviation in Australia
(b) To establish a system for the regulation of airports that has due regard to the interests of airport users and the general community
(c) To promote the efficient and economic development and operation of airports
(d) To facilitate the comparison of airport performance in a transparent manner
(e) To ensure majority Australian ownership of airports
(f) To limit the ownership of certain airports by airlines
(g) To ensure diversity of ownership and control of certain major airports
(h) To implement international obligations relating to airports.

6.1.1 Policy

Countries transfer the responsibility for development and management of airports to the private sector under the conviction that such activities should not be the responsibility of the government. Such has been the case in the UK; the government, through its Airports Act of 1986, dissolved the British Airports Authority (BAA) and vested its property in a successor private company: BAA plc (later to be broken up into individual airports).

The Canadian Federal Government defined a new policy in 1987 called *A Future Framework for Airports in Canada*, to implement the transfer of most airports to regional, provincial or local authorities to manage and operate airports through long term ground leases with Transport Canada. In 1992, not-for-profit local airport authorities were given the mandate to operate airports in Montreal, Vancouver, Calgary and Edmonton. In 1994, the Federal Government, through the National Aviation Policy (NAP), sold small and regional airports to their communities at a nominal price.[1] Airports with scheduled traffic of more than 200,000 annual passengers, considered nationally significant, constitute the National Airport System (NAS) including 26 airports accounting for 94% of the overall traffic. For this group of airports, the NAP states:

> The development of the NAS reflects a commitment on the part of the federal government to the viability of a national system of safe, commercially-oriented and cost-effective airports. While the federal government will guarantee the ongoing viability of the NAS as a whole, this does not necessarily mean its continued direct operation or funding. The government will commercialize the NAS through the transfer of responsibility for the operation, management and development of NAS airports to Canadian Airport Authorities. Local operation is preferable since it has proven to be more cost-effective, more responsive to local needs and better able to match levels of service to local demands.
>
> As a general rule, airports within the NAS will be required to become financially self-sufficient (operating and capital costs) within five years beginning April 1, 1995. For certain NAS airports, it is recognized that undercapitalization in the past or future capital requirements may result in some adjustments to this principle.[2]

Through this, Transport Canada transferred the responsibility for managing airports under commercial principles from the federal government to local governments.

The New Zealand government also transferred the operation of their two largest airports, Auckland and Wellington, to the private sector. In 1985, during parliamentary debates, the Minister of Transport stated that airports were to be a commercially viable concern and investments should be financed by airport revenue.[3] By the 1990s, Auckland airport was listed on the New Zealand Stock Exchange and on the Australian Stock Exchange.

The governments of the UK, Canada, Australia and New Zealand considered as a matter of principle that the operation and investment in airports should be the responsibility of the private sector, regardless of the availability of public funds.

---

1 Jed Chong, "Airport Governance Reform in Canada and Abroad" (2017).
2 Government of Canada, National Aviation Policy (2010).
3 See the New Zealand Parliament's *Parliamentary Debates. House of Representatives*, Volume 466 (1985).

6.1.2 Private financing of investments in infrastructure,

For many countries, the need for financing expansion or improvements outside of the state treasury has been the main driver for PPP.

As explained in Chapter 5, airports have to comply with a specific set of regulations, including those defined by ICAO as the Standards and Recommended Practices (SARPs) and adapted in most countries[4] through local regulations. In addition, airports seek to comply with specific levels of service to passengers and aircraft operators (including airlines), often defined by international bodies like IATA or ACI.

In many countries in the developing world, governments face the need to update their facilities to comply with these standards, by adapting to new requirements or by simply upgrading the facilities, in many cases after long periods of neglected maintenance. To provide a better level of service by expanding and enhancing the facilities, significant investments are required.

Financing capex from the state budget can be challenging, given competing needs in other sectors like health, education, public services, national security, etc. Airports are often perceived as services for the rich (despite their benefits for the economy and population at large) and therefore may attract less public financing.

This happened in Madagascar for example, where the airports operated by the state owned enterprise ADEMA (Aéroports de Madagascar) were in serious need of upgrading and expanding. As an initial stage, in 2017, the government transferred the two largest airports, Antananarivo Ivato and Nosy Be Fascene, to a consortium of French companies for a period of 28 years.[5] During 2018 both airports were upgraded significantly, as part of a €215m investment program that included a new terminal building for Ivato, major refurbishment and strengthening of the runways, and apron expansion.

Almost all airports transferred to the private sector in Latin America since the early 1990s were in need of significant works of improvement; there was a need to adapt to international norms of safety and security, including major repairs for obsolete infrastructure. In addition, the necessity to accommodate traffic with appropriate levels of service and comfort was evident under the public administration. Such has been the case of the dramatic transformation of airports such as Arturo Merino Benítez in Santiago, Jorge Chávez in Lima, El Dorado in Bogotá, and Carrasco in Montevideo, to name a few. Governments have been unable to raise the funds required to improve these facilities, and have turned to PPP to mobilise financing.

---

4 All signatory countries of the Chicago Convention of 1944 are automatically obliged to homologate the SARPs under their own regulations.
5 Aeroports de Paris (ADPm) as the operator, Bouygues Bâtiment International, Colas Madagascar and Meridiam, forming Ravinala Airports.

6.1.3 Improving management of airport

For most PPPs in the developing world, the motive of financing infrastructure was combined with the need to improve the management of the airport. Convinced that private operators could provide better levels of service and achieve greater efficiencies in management and construction, governments considered partnerships with these operators and developers.

For example, the Egyptian Holding Company for Airport & Air Navigation (EHCAAN) engaged Frankfurt Airport (Fraport) through a management contract to operate Cairo Airport for a period of ten years. The government expanded the airport with their own resources, but brought in Fraport to improve the operations.

During the first round of concessions in the 1990s in Chile, the government focused on financing but not on management. The Ministry of Public Works (MOP) transferred, sequentially, eleven airports using a concession model for relatively short periods of time. The tender conditions (called the BALIs) were in fact more appropriate for a construction concession than for management. The government of Chile was looking for the private sector to develop the facilities but were expecting to get them back as soon as the operators could recuperate their investment, under the belief that the public sector could manage the airports as well as the private sector. However, the improved management that resulted from the first wave of concessions made the MOP reconsider the concession terms when defining the conditions for the second round of concessions. The concessions designed after 2012 included an extended concession period of 20 years, and incorporated a set of indicators and parameters significantly more focused on operations and management.

The International Air Transport Association (IATA) has developed an Airport Design Reference Manual (ADRM) that defines a world standard for level of service. Formulated through very objective indicators for each of the subsystems of an airport, IATA's LoS is probably the most reliable measurement of quality of service for a terminal building. Awards and qualifications from other international rating agencies like Skytrax or Airports Council International (ACI) are also relevant indicators of airport management.

Most importantly, the level of compliance of an airport with ICAO's standards and recommended practices (SARPs) with respect to safety and security is the single most important indicator of good performance. These standards are the minimum acceptable indicators to be expected under any management. No PPP should be considered successful if it breaches compliance with the SARPs.

Practically every airport PPP around the world enhanced commercial exploitation; the amount, variety and quality of commercial services, including duty-free shopping and travel retail, food and beverage, general retailing and transportation services, to name the most relevant, improved significantly.

Governments have sustained the benefit of PPP reform by setting higher levels of service (LoS). For some countries the situation before the PPP was so tremendously bad that new management, even if inefficient, generated a very positive impact on passengers.

The level of sophistication of passengers varies across different regions, depending on income level, culture and habits. Therefore, an airport might provide good service from the perspective of some passengers and unacceptable for others.

6.2 Measuring the success of PPP

The success of PPP reflects the motivation of governments in adopting it, but should fundamentally focus on the extent to which the private sector delivers better infrastructure in the most efficient way, achieving compliance with the highest standards of safety and security while providing better levels of service to users, without a significant increase of cost to the travelling public and airlines. The PPP agreement will establish a performance regime with minimum technical requirements (MTRs) for reporting to the contracting authority, providing an important input for its monitoring of operational performance. See Chapter 10 for further discussion of monitoring regimes. Some MTRs will be associated with penalty regimes (and possibly bonuses) ensuring the project company will achieve the government's service levels.

Measuring success involves a comparison of the situation with and without private sector participation. Measuring the performance of state-owned airports (SOE) as opposed to airports operated by the private sector may provide the answer. The comparison can be carried out by measuring financial performance with four types of key performance indicators (KPIs):

1. Total revenues per passenger
2. Commercial revenues per passenger
3. Operating expenses per passenger
4. Operating expenses per air traffic movements (ATM).

The graph in Figure 6.1 measures revenues collected by the airport (aeronautical and non-aeronautical) per unit of throughput, particularly the number of passengers.

The comparison of total revenues per passenger suggests that airports under PPPs have a better performance than airports operated by the public sector. The sample average for the former is roughly 25% higher than that of the latter. However, it is important to note that the component of aeronautical revenues – which typically account for between 60% and 70% of the total revenues – is not necessarily related to good performance of the operator. Fees and charges (on airlines and on passengers) are typically regulated and imposed by the government. An airport with high fees and charges does not mean that it is better managed.

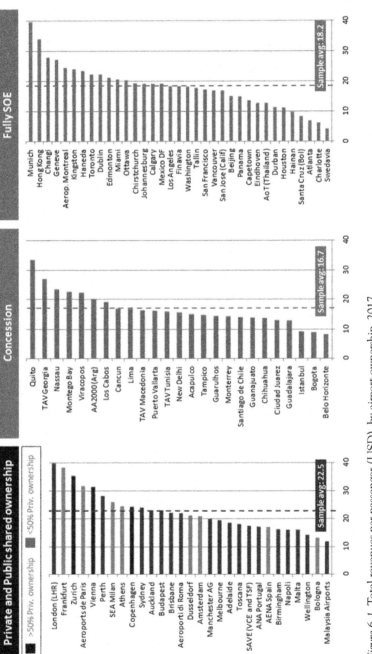

Figure 6.1 Total revenues per passenger (USD), by airport ownership, 2017
*: Data for individual airports or airports holding companies when non-differentiated
Source: Authors, 2017 data

Assuming that better management is related to a higher yield from commercial activities, a comparison of the commercial revenues per passenger is shown in Figure 6.2.

This analysis suggests that airports under some sort of shared ownership tend to outperform both public airports as well as airports under concession schemes. However, the differences could be attributable to other reasons. Airports under shared ownership are typically found in developed economies, where the purchasing power of the traveller is significantly higher than in airports in the developing world. Concession models are commonly found across Latin America and India, where the purchasing power and the currency rate against the US result in significantly lower dollar denominated revenues per passenger.

The good results for airports operated by the public sector may be because publicly run airports for which information is available through published, transparent accounting tend to be professionally managed and thus, better performers. The opposite is also true: poorly managed public sector airports often do not publish their results. SOE airports that publish their financial results are mostly from developed economies where the public sector has better capacity. These countries, in Europe, Australia, New Zealand and North America, enjoy higher commercial consumption rates per passenger, with opportunities for commercial activities. The list includes a number of Asian airports (Singapore, Tokyo Haneda, Hong Kong and Beijing) which typically present the highest consumption patterns, particularly by sales of luxury brands and high-end duty-free shops.

When measuring operating expenses (opex) per unit of throughput of passengers or of air traffic movement, the results are not conclusive.

The comparison of operating expenses per passenger across different ownership schemes shows that airports under concession have a lower sample average than airports with shared ownership; it is also lower than that of state-owned airports. The fact that most of the airports within the sample of concessions are in Latin America and India explains the lower operating costs, particularly due to the lower cost of labour and lower levels of costs in dollar terms.

In addition, as discussed, the fact that those SOE airports that publish their results tend to be the better managed ones also distorts the results. Information on most publicly run airports that are not efficiently managed is not easily available.

Similar results were obtained when comparing expenses in relation to ATMs (see Figure 6.4).

Airports under concession arrangements, in this particular sample, present an average which is 30% lower than that of airports operated by SOEs and less than half of the sample average corresponding to privatised airports. As with the results above, lower production costs in India and Latin American countries and the availability of information for only the better operated SOE airports may distort this analysis.

Evidence of the impact and success of PPP reform should include the performance of the airport before and after the reform in terms of financial results. However, measuring *before* and *after* may not always be so straightforward. Historic financial information of airports previously owned or operated by SOEs

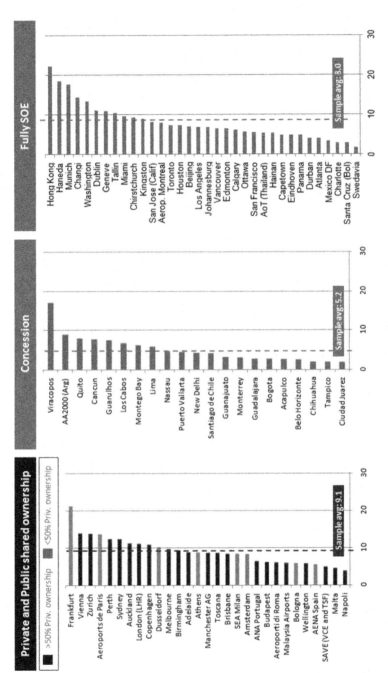

Figure 6.2 Commercial revenues per passenger (USD), by airport ownership, 2017
*: Data for individual airports or airports holding companies when non-differentiated

Source: Own research, 2017 data

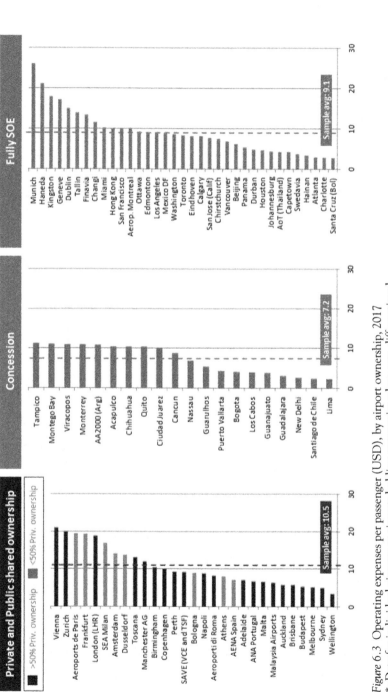

Figure 6.3 Operating expenses per passenger (USD), by airport ownership, 2017
*: Data for individual airports or airports holding companies when non-differentiated
Source: Authors, 2017 data

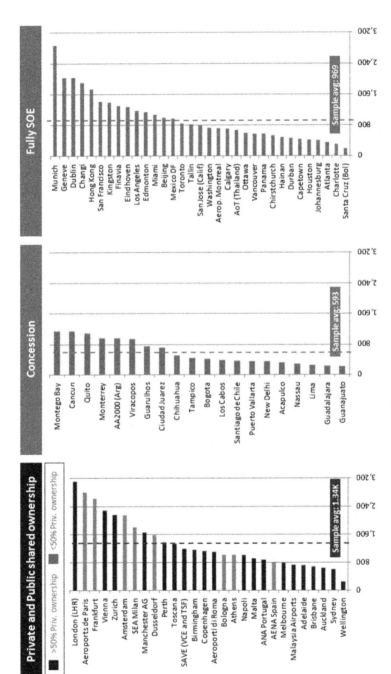

Figure 6.4 Operating expenses per ATM (USD), by airport ownership
*: Data for individual airports or airports holding companies when non-differentiated
Source: Authors, data from end FY2017

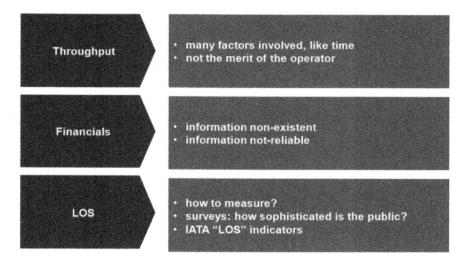

Figure 6.5 Measuring performance "before" and "after"

is rarely available, making it very difficult to obtain the necessary information for the *ex-ante* analysis. Most publicly owned airports do not publish their financials, because of lack of transparency or because they present their financial statements in an aggregated manner with other government agencies. In other countries, the historical financial information may be available but might be unreliable, making it difficult to compare against other benchmarks or even to measure their own evolution.

6.3 Different forms of airport PPP

The scope of an airport PPP is defined by the services to be included in the contract. An airport can be divided in two areas: airside and landside. The airside includes all the operational facilities in the airfield, including the runway system, the taxiways, the apron and all the associated support areas. The landside comprises the passenger terminal building(s), the cargo terminal building(s) and all the access areas, including the car parking facilities.

Some airport PPPs involve the private sector investing in either the landside only, or the entire airport. A PPP for the airside only is very rare. The second runway for Bogotá El Dorado airport was developed under a PPP; the investor recuperated its capital expenditures in the new runway through the landing fees collected at both runways, less their associated maintenance costs.[6]

---

6 The concession for the second runway, operated by CODAD, expired in September 2017 and the operation of both runways reverted to the Civil Aviation Authority of Colombia (Aerocivil).

112 *Airport PPPs around the world*

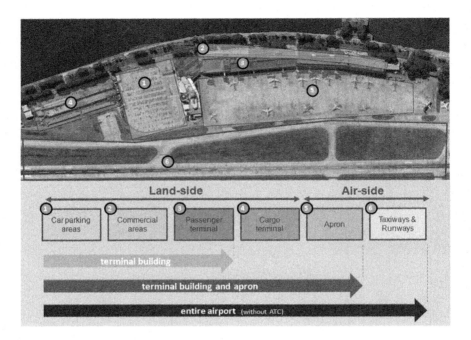

Figure 6.6 Alternative scope of an airport PPP

PPP schemes for only the landside are more common. For example, concession contracts exclusively for the construction, operation and maintenance of the terminal building are found at Liberia Airport in Costa Rica, Santiago de Chile, Orlando Sanford and JFK's Terminal 4. Particularly in the US, it is common to find long-term lease arrangement with airlines, where carriers assume the responsibility to build and operate parts of entire terminal buildings.

Once scope is defined, the contracting authority will need to define the PPP model that fits best with requirements. PPP models are ultimately flexible[7] to fit with Government needs. This book will explore three basic models, for ease of reference, and which reflect the most common forms of airport PPP:

1. Divestiture (share or equity sale)
2. Concession models (including BOT-type)
3. Management contract.

Alternative models for private sector participation in airport operations range from schemes with minimal private sector ownership, such as management

---

7 For more, see Delmon's working paper for the World Bank, "Understanding Options for Public Private Partnerships in Infrastructures" (2010).

contracts or operating concessions, to those where ownership is fully transferred, such as a share sale. In between are other models such as Build-Operate-Transfer (BOT) and Build-Own-Operate-Transfer (BOOT).

PPP models have specific characteristics:

Ownership: in a sale of shares or equity, partial or complete ownership is transferred to the private sector indefinitely. A concession/BOT contract transfers, sometimes with ownership, the right to operate the assets to the private sector for a limited time. Management contracts maintain the ownership within government for the whole contract duration.

Risk transfer: PPP projects typically transfer financial risks to the private sector, allowing airport performance improvement and construction efficiency. The more ownership control is given away, the more risk can be transferred to the private sector. A management contract will transfer limited risk to the private sector; share or equity sale typically transfers maximum risk.

Contract duration: typically management contracts are for five years, concessions for a longer period (often 25+ years), and divestitures are considered permanent.

Therefore, choosing the form of private participation is a compromise between ownership, degree of risk transfer and contract duration. Hence, the shorter the contract duration, the more difficult it will be to attract private investment in airports.

6.3.1 Divestiture/Equity sale

Equity sale generally involves transferring airport property to a private sector entity through lease or outright sale, meaning the transfer is considered permanent. These models are commonly found in Anglo-Saxon countries, in the form of long-term leases (e.g. 50 + 50 years). Such is the case in the UK (Heathrow, Gatwick, etc.), and in the Australian and New Zealand airports (Sydney, Melbourne, Auckland, Wellington, etc.). This form of PPP is also quite commonly found at many European airports (Portugal, Germany, the Netherlands, etc.).

Equity sales can involve "corporatising" the airport and subsequent sale of significant shareholdings in such an entity to the private sector (company/ies or through the stock exchange). Under this arrangement, government sells equity in a corporatised airport authority to the private sector for injection of capital and management expertise. The airport corporation owns the airport property and/or rights to operate it, functioning like a commercial entity under a mixed public-private ownership. Corporatising is not a guarantee of success. Like any company, the corporate entity must be profitable and have incentives to perform its functions efficiently.

The number of shares sold (under private sector control) is not merely a financial transaction. If a private operator owns the shares of an airport, it should be allowed to make changes and improvements, subject to legal and regulatory

114 *Airport PPPs around the world*

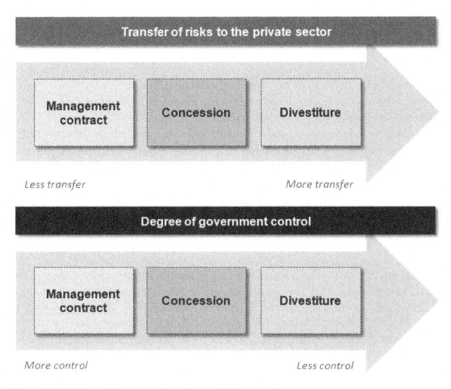

Figure 6.7 Alternative scope of an airport PPP

requirements. The organisational structure, the freedom to hire and fire staff and the ability to make contractual and investment decisions are essential for the successful operation of airports. An appropriate governance structure should be established and audited financial statements should be published regularly; both of these require reforms prior to seeking private participation.

Box 6.2 Selling the family jewels

Airports are high value assets. During difficult economic times, decision makers may be tempted to privatise airport as a way to release value (i.e. cash) for other purposes. Selling the family jewels to cover short-term needs can create numerous problems, in particular where the purpose of privatisation is not linked to improving and expanding services. Often, the project contracts are not geared towards proper operation of the airport. Also, during crisis, airports selected are typically not well performing airports, resulting in a discounted price. It may make more financial sense to expand or redevelop the airport first, before privatisation.

The government may prefer to sell down a minority equity stake in the airport company, for example through private or public offering. This may be a first step toward improving governance of the company and share value. When the UK decided to divest shares in BAA in 1987, it started with a minority equity sale. In 2006, the French government sold about 30% of Aéroports de Paris (ADP) through an IPO.[8]

Majority equity sale or full divestment involves government giving up control and ownership to investors. Returns from such divested assets are higher given the ownership premium. Investors are likely to pay much less for a minority equity sale. Other mechanisms, such as golden shares, may be used to protect government interest despite divestiture. The public offering of BAA shares on the London stock exchange in 1987 is an early example of airport divestment. The government retained a "golden share" with the primary intention of preventing takeover by foreign investors. This practice ceased in 2003, under EU law.

A typical challenge is that the legal regime often forbids the sale of shares due to the strategic responsibility assigned to airports. This is usually addressed during the feasibility stage (see Chapter 8). Often there is resistance to selling shares, as sales of state assets raise concerns about fair price and potential lay-off of staff, or illustrate mistrust of the private sector.

6.3.2 Concession and Build-Operate-Transfer (BOT)

Operating concessions may involve a significant part of the airport; this may include the passenger terminal building and/or the landside areas, or the total airport property. The airport owner signs a contract with a specialised operator to manage, operate, and maintain the property. The contractor is expected to manage and cover all operating costs with revenue generated by aeronautical and non-aeronautical airport activities included in the concession. The government remains the owner of the fixed airport property and is responsible for defining and contributing, if needed, to capital investment; the concessionaire is usually responsible for arranging financing and undertaking the airport construction/improvement, and will own movable property, like vehicles.

Government receives a fee from the concessionaire for the right to operate the airport. Fees are typically a fixed amount (eventually as an upfront payment), a variable amount based on total revenues or on a per passenger basis, or a combination of both. Potential revenue from concession fees is determined by a financial assessment during the feasibility phase. The assessment will determine if government contribution is needed to make the project financially viable for a concessionaire. The government's actual contribution is confirmed during the bidding phase and may be part of the selection criteria.

The mechanism for such BOT arrangements is generally a medium/long-term lease contract. The government retains airport property ownership and the

---

8 The French government, as of August 2019, retains 50.6% of the capital shares of ADP.

Table 6.1 Examples of share sales or capitalisation of airports

Country	Cities	Airports
Australia	Sydney, Melbourne, Adelaide, Perth and another 17 cities	Kingsford Smith, Tullamarine, Adelaide International, Perth International and another 17 airports
Austria	Vienna	Vienna International
Belgium	Brussels, Antwerp	Brussels National, Antwerp International
Denmark	Copenhagen	Copenhagen Airport Kastrup
France	Paris, Toulouse, Nice, Cannes, Saint Tropez, Lyon, Rennes, Nantes, Grenoble	Charles de Gaulle, Orly, Toulouse Airport, Cote D'Azur, Cannes-Mandelieu, La Mole, Saint Exupéry, Saint-Jacques, Atlantique, Isere
Germany	Munich, Frankfurt, Memmingen	Franz Josef Strauss, Frankfurt am Main, Allgau
Hungary	Budapest	Budapest-Ferenc Liszt
Italy	Rome, Florence, Naples, Parma, Turin, Treviso, Milan, Bologna	Fiumicino, Ciampino, Peretola, Capodichino, Parma Airport, Caselle, Sant'Angelo, Linate, Malpensa, Marconi
Malaysia	Kuala Lumpur and another 37 cities	Kuala Lumpur International, Kota Kinabalu, Penang, Kuching, Senai and another 32 airports
Malta	Malta	Malta International Airport
Moldavia	Chisinau	Chisinau Airport
Netherlands	Amsterdam	Schiphol Airport
New Zealand	Auckland and Wellington	Auckland International and Wellington International
Poland	Katowice	Międzynarodowy Port Lotniczy
Portugal	Lisbon, Oporto and another 8	Portela, Sá Carneiro, Faro, Joao Paulo II, Beja, Cristiano Ronaldo, Porto Santo, Azores, Horta, Das Flores
Russia	Moscow	Domodedovo, Sheremetyevo, Vnukovo
Slovenia	Ljubljana, Maribor	Jože Pučnik Airport, Maribor Airport
Spain	Barcelona, Madrid, Mallorca, Seville, Valencia, another 43	Barajas, El Prat and another 46 airports
Switzerland	Zurich	Zurich Kloten
UK	London, Manchester, Derby, Liverpool, Aberdeen, Glasgow, Edinburgh	Heathrow, Stansted, Gatwick, London City, Luton, East Midlands, John Lennon, Aberdeen, Glasgow, Edinburgh
Ukraine	Odessa, Kharkiv	Odessa Airport, Kharkiv Airport

Source: Authors' research, June 2019
Note: at least 5% of private shareholding

concessionaire pays the owner a concession fee that could be a fixed amount, a percentage of revenues or a combined formula that distributes risk between government and private sector. The concessionaire is typically responsible for financing airport operations and in most cases, for making investment and improvements according to a specified schedule of works. Profits and losses accrue to the concessionaire.

This model has been widely adopted in the developing world, particularly throughout Latin America, India and Africa. Its success has been limited in countries with weak institutional and regulatory frameworks because some investors have not complied with their original commitments in terms of investments or costs to users. Chile is among the most successful examples of this model, where eleven airports have been subsequently awarded as individual concessions. All have significantly upgraded their infrastructure with minimal or no additional cost to users.

The Build, Own, Operate and Transfer (BOOT) model is a variant in which asset ownership is temporarily transferred to the concessionaire during the term of the contract. A BOOT model allows the private sector to pledge the transferred assets to secure financing for development, facilitating the financing scheme, and thus making it more appealing for investors.

6.3.3 *Management contract*

This is one of the simplest forms of private sector participation; the private sector is paid to perform specific functions and deliver specific services. Management contracts generally do not require the private sector to invest or develop commercial activities. It does not transfer much risk to the private sector but has specific advantages.

Management contracts are easier to prepare than a concession/BOT and do not require extensive preparation or transaction support. The contract duration is short, usually five years, and is attractive for a government with limited experience in private sector participation. Service quality and efficiency improvement can be achieved through a management contract. As a result, a management contract can be a first step towards a private sector participation strategy. The main risk is that a single operator gains valuable knowledge about the airport operations, which provides it with an advantage in future bidding.

6.4 Geographical distribution and implementation of airport PPPs

According to Airports Council International, global passenger traffic reached 8.3 billion in 2017. Out of this number, approximately 27% – roughly over 2 billion passengers – passed through airports operated under some kind of PPP arrangement.

In 2017, airports operated under PPP concession arrangements represented 9.5% of the ACI airports' total passenger throughput. Another 17.3% of

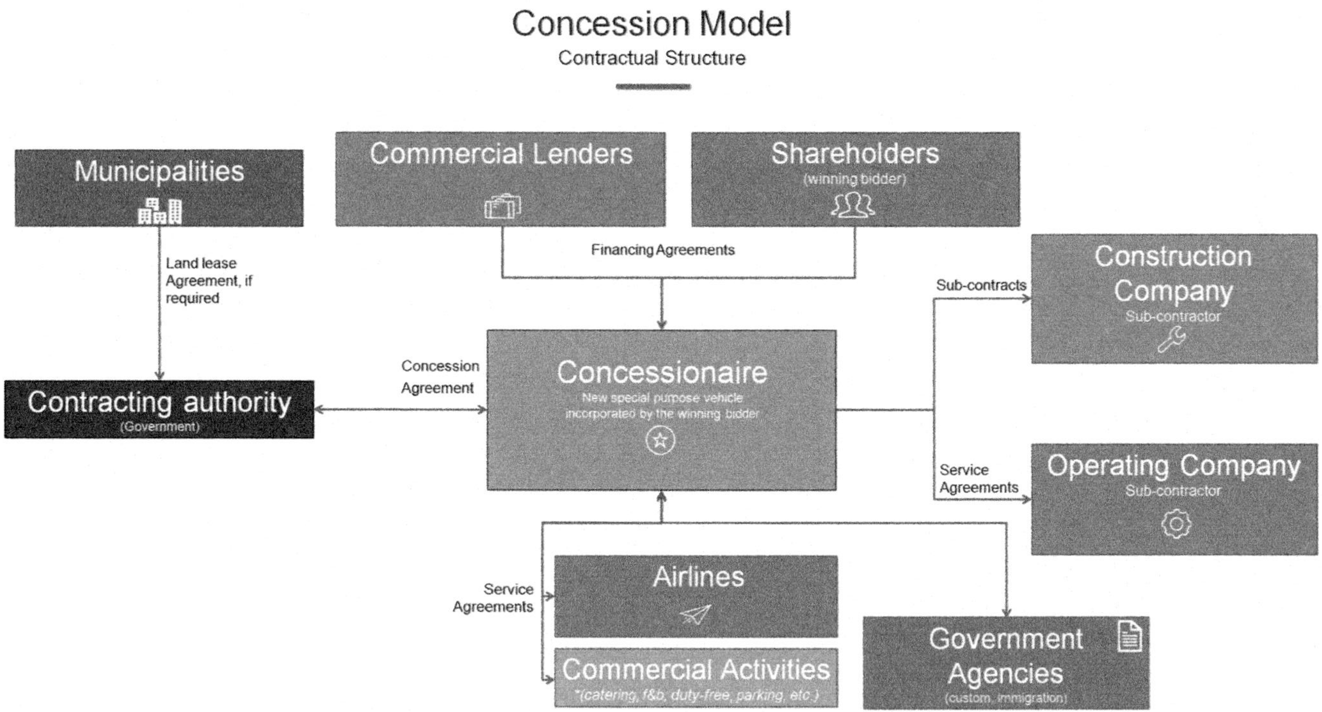

Figure 6.8 Contractual structure of a concession model
Source: Adele Paris, IFC

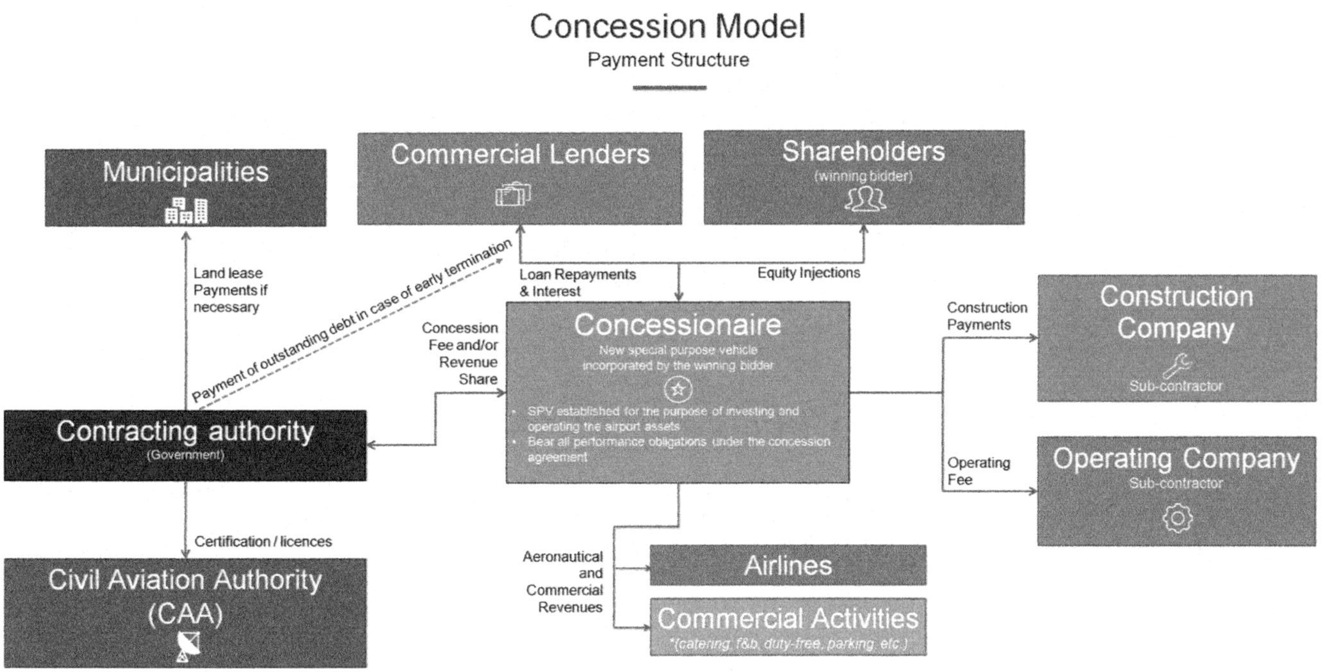

Figure 6.9 Payment structure of a concession model

Source: Adele Paris, IFC

Table 6.2 Examples of concessions

Country	Cities	Airports
Albania	Tirana	Nënë Tereza
Argentina	Buenos Aires, Cordoba, Mendoza, Salta, Bariloche and another 30 cities	Ezeiza, Aeroparque, El Plumerillo, Neuquen, San Fernando, El Palomar, Ing. Ambrosio Tavella, Mar del Plata and another 28 airports
Armenia	Yerevan, Shirak	Zvartnots, Shirak International
Brazil	Sao Paulo, Brasilia, Rio de Janeiro, Natal, Fortaleza, Belo Horizonte, Florianopolis, Porto Alegre, Salvador	Guarulhos, Viracopos, Galeao, Grande Natal, Presidente Juscelino Kubitschek, Hercílio Luz, Confins, Salgado Filho, Deputado Magalhães, Pinto Martins
Bulgaria	Burgas, Varna	Burgas Airport, Varna Airport
Cambodia	Siem Reap, Sihanoukville and Phnom Penh	Siem Reap, Sihanoukville and Phnom Penh
Chile	Santiago, Arica, Iquique, Antofagasta, Calama, Atacama, La Serena, Concepción, Temuco, Puerto Montt, Punta Arenas	Arturo Merino Benítez, Chacalluta, Diego Aracena, Andres Sabella, El Loa, Atacama, La Florida, Carriel Sur, La Araucania, El Tepual, Carlos Ibáñez del Campo
Colombia	Bogotá, Cali, Barranquilla, Santa Marta, Cartagena, Medellín and another 11 cities	El Dorado, Bonilla Aragon, Ernesto Cortíssoz, Simon Bolivar, Rafael Nuñez, Jose Maria Cordova and another 11 airports
Congo	Brazzaville, Pointe-Noire and Ollombo	Maya Maya, Agostinho Neto and Ollombo airports
Costa Rica	San Jose and Liberia	Juan Santamaria and Ernesto Cortíssoz
Cote D'Ivoire	Abidjan	Port Bouet Airport
Croatia	Zagreb	Franjo Tudman
Cyprus	Paphos, Larnaca	Paphos International, Larnaca International
Dominican Republic	Santo Domingo, Puerto Plata, Barahona, Samana	Las Americas, Joaquín Balaguer, Gregorio Luperón, Maria Montez, Juan Bosch, Arroyo Barril
Ecuador	Guayaquil, Quito, Galapagos	Jose Joaquin de Olmedo, Mariscal Sucre, Isla Baltra
French Polynesia	Tahiti, Bora Bora, Raiatea and Rangiroa	Tahiti Fa'a'a, Bora Bora, Uturoa and Rangiroa
Gabon	Libreville	Leon M'ba

Table 6.2 (Cont.)

Country	Cities	Airports
Georgia	Tbilisi, Batumi	Novo Alexeyevka, Batumi airport
Greece	Athens and 14 regional airports	E. Venizelos, Mykonos, Santorini and another 11 airports
Honduras	Tegucigalpa, San Pedro Sula, Roatan y La Ceiba	Toncontin, Ramon Villeda Morales, Juan Manuel Galvez, Goloson
India	New Delhi, Mumbai, Hyderabad and Bangalore	Chhatrapati Shivaji, Indira Gandhi, Rajiv Gandhi and Kempegowda
Jamaica	Kingston and Montego Bay	Norman Manley International and Sangster International
Japan	Kansai, Osaka and Sendai	Kokusai, Itami and Sendai
Jordan	Amman	Queen Alia International Airport
Kosovo	Pristina	Pristina Airport
Macedonia	Skopje, Orhid	Alexander the Great, St Paul the Apostle
Madagascar	Antananarivo and Nosy Be	Ivato and Fascene airports
Mexico	Cancun, Monterrey, Guadalajara and another 31 cities	Cancun International, General Mariano Escobedo, Miguel Hidalgo y Costilla and another 31 airports
Peru	Lima and another 17 cities	Jorge Chávez International and another 17 regional airports
Philippines	Mactan Cebu and Clark	Clark International and Mactan Cebu International
Puerto Rico	San Juan	Luis Muñoz Marín
Russia	St Petersburg	Pulkovo
Saudi Arabia	Madinah	Prince Mohammad Bin Abdulaziz Airport
Serbia	Belgrade	Nikola Tesla
Tunisia	Hammamet and Monastir	Enfidha and Habib Bourguiba airports
Turkey	Istanbul, Antalya, Ankara, Izmir, Gazipasa, Milas	Antalya Airport, İstanbul New airport, Sabiha Gökçen, Esenboğa Airport, Adnan Menderes, Alanya, Bodrum
Uruguay	Montevideo and Punta del Este	Gral. Cesáreo L. Berisso and Laguna del Sauce

Source: Authors' research, June 2019

122 Airport PPPs around the world

Table 6.3 Examples of airports management contracts

Country	Cities	Airports
Bahamas	Nassau	Linden Pyndling International
Egypt (expired)	Cairo	Cairo International
Saudi Arabia (expired)	Riyadh, Jeddah	King Khalid International, King Abdulaziz International

Source: Authors' research, June 2019

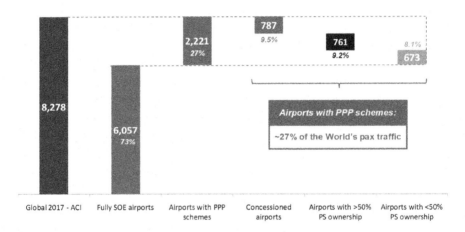

Figure 6.10 Passenger traffic through airports by type of ownership, 2017 in millions of pax (approximate numbers)
Source: Author research and ACI 2017 data

passengers travel through airports with some type of private shareholding. The distribution is about equal between passengers that passed through airports with a majority private shareholding and those who passed through an airport with a minority private ownership.

When looking at the geographical distribution of the different PPP models, the concession model is particularly popular in developing economies; it is extensively used in Latin America, in Asia and in some countries in Africa. In Europe, both concessions and share sale models are found.

In 2017, there were approximately 176 airports in the world that had been privatised through share sales across 21 countries, while roughly 194 airports were operated under 73 concession contracts.

Although passenger traffic at airports operated by SOEs still represents a significant portion of the total passenger traffic in Europe, privatisations through share sales of AENA in Spain, of ANA in Portugal, of the largest airports in the

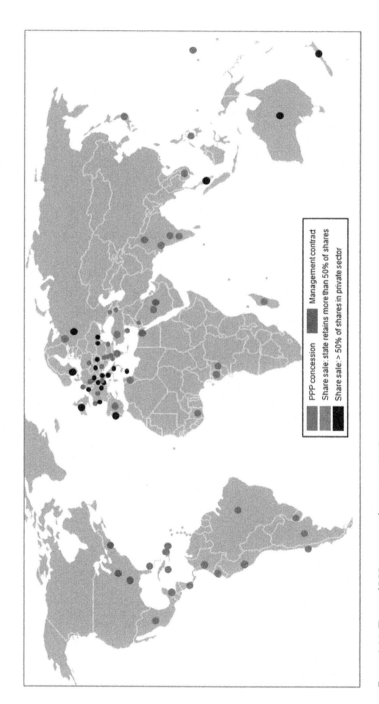

Figure 6.11 Types of PPP contracts by region, 2017
Source: Author research, 2017 data

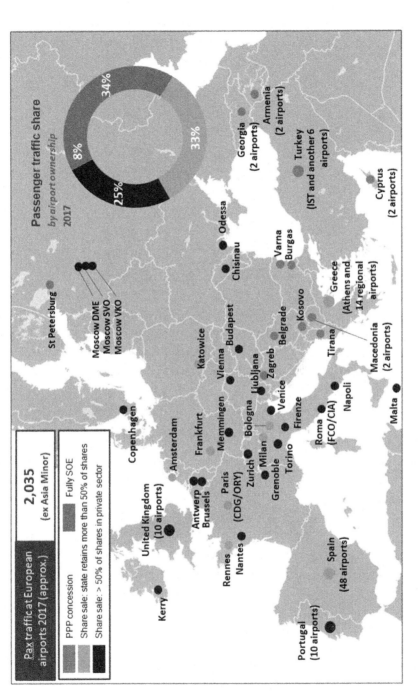

Figure 6.12 PPP airports in Europe and passenger traffic share by airport ownership, 2017

Source: Author research, 2017 data

Table 6.4 Airports PPP in Europe

Country	Cities	Airports	PPP Scheme*
Albania	Tirana	Nënë Tereza	C
Armenia	Yerevan, Shirak	Zvartnots, Shirak International	C
Austria	Vienna	Vienna International	SS
Belgium	Brussels, Antwerp	Bruxelles-National, Antwerp International	SS
Bulgaria	Burgas, Varna	Burgas Airport, Varna Airport	C
Croatia	Zagreb	Franjo Tuđman	C
Cyprus	Paphos, Larnaca	Paphos International, Larnaca International	C
Denmark	Copenhagen	Copenhagen Airport Kastrup	SS
France	Paris, Toulouse, Nice, Cannes, Saint Tropez, Lyon, Rennes, Nantes, Grenoble	Charles de Gaulle, Orly, Toulouse Airport, Cote D'Azur, Cannes-Mandelieu, La Mole, Saint Exupéry, Saint-Jacques, Atlantique, Isere	SS
Georgia	Tbilisi, Batumi	Novo Alexeyevka, Batumi airport	C
Germany	Munich, Frankfurt, Memmingen	Franz Josef Strauss, Frankfurt am Main, Allgau	SS
Greece	Athens and 14 regional airports	E. Venizelos, Mykonos, Santorini and another 11 airports	C
Hungary	Budapest	Budapest-Ferenc Liszt	SS
Italy	Rome, Florence, Naples, Parma, Turin, Treviso, Milan, Bologna	Fiumicino, Ciampino, Peretola, Capodichino, Parma Airport, Caselle, Sant'Angelo, Linate, Malpensa, Marconi	SS
Kosovo	Pristina	Pristina Airport	C
Latvia	Riga	Riga Airport	MC
Malta	Malta	Malta International Airport	SS
Macedonia	Skopje, Orhid	Alexander the Great, St Paul the Apostle	C
Moldavia	Chisinau	Chisinau Airport	SS
Netherlands	Amsterdam	Schiphol Airport	SS
Poland	Katowice	Międzynarodowy Port Lotniczy	SS
Portugal	Lisbon, Oporto and another 8	Portela, Sá Carneiro, Faro, Joao Paulo II, Beja, Cristiano Ronaldo, Porto Santo, Azores, Horta, Das Flores	SS
Russia	Moscow	Domodedovo, Sheremetyevo, Vnukovo	SS

(continued)

Table 6.4 (Cont.)

Country	Cities	Airports	PPP Scheme*
Russia	St Petersburg	Pulkovo	C
Serbia	Belgrade	Nikola Tesla	C
Slovenia	Ljubljana, Maribor	Jože Pučnik Airport, Maribor Airport	SS
Spain	Barcelona, Madrid, Mallorca, Seville, Valencia, another 43	Barajas, El Prat and another 46 airports	SS
Switzerland	Zurich	Zurich Kloten	SS
Turkey	Istanbul, Antalya, Ankara, Izmir, Gazipasa, Milas	Antalya Airport, İstanbul New airport, Sabiha Gökçen, Esenboğa Airport, Adnan Menderes, Alanya, Bodrum	C
Ukraine	Odessa, Kharkiv	Odessa Airport, Kharkiv Airport	SS
UK	London, Manchester, Derby, Liverpool, Aberdeen, Glasgow, Edinburgh	Heathrow, Stansted, Gatwick, London City, Luton, East Midlands, John Lennon, Aberdeen, Glasgow, Edinburgh	SS

Source: Author research, 2017 data
*: C = Concession; SS = Privatisation/Share Sale; MC = Management Contract

UK and several others in the past 30 years have transformed the landscape of airport ownership and management in this region.

In 2017, airports with some degree of private ownership through share sales represented approximately 58% of the passenger traffic processed at European airports, estimated at roughly 2,035 billion passengers. Airport PPP contracts as concessions have been implemented in the past 15 years in Europe, with St Petersburg, Athens, Istanbul,[9] the Greek islands, Zagreb, Belgrade, Tirana, Varna, Burgas, Istanbul, Skopje, Tbilisi, Batumi, Yerevan and another six airports in Turkey[10] being the most relevant examples. This type of PPP arrangement represented roughly 8% of the region's traffic in 2017.

---

9 The concession of Istanbul New Airport (which started operating in April 2019) has been awarded to a consortium of five Turkish companies. Before the opening of the new airport, Ataturk International Airport was the city's most important airport, and it was concessioned to and operated by TAV Airports.
10 Ankara, Izmir, Alanya, Bodrum and Antalya airports (concessioned to TAV airports), along with Sabiha Gocken airport (concessioned to a joint consortium composed of Malaysia Airports and TAV airports).

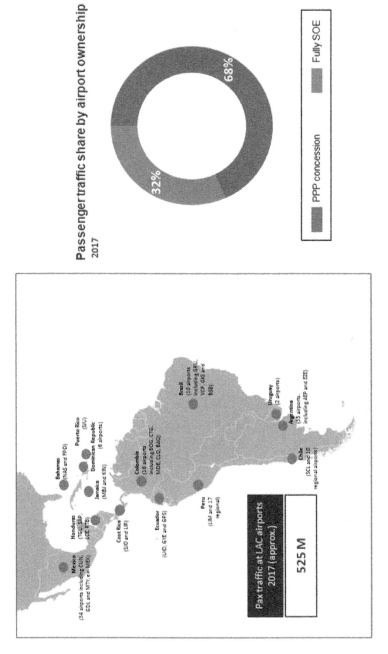

Figure 6.13 PPP airports in Latin America and the Caribbean and passenger traffic share by airport ownership, 2017

Source: Author research, 2017 data

Table 6.5 Airport PPPs in Latin America

Country	Cities	Airports	PPP Scheme*
Argentina	Buenos Aires, Cordoba, Mendoza, Salta, Bariloche and another 30 cities	Ezeiza, Aeroparque, El Plumerillo, Neuquen, San Fernando, El Palomar, Ing. Ambrosio Tavella, Mar del Plata and another 28 airports	C
Bahamas	Nassau and Freeport	Linden Pyndling and Grand Bahama	C
Brazil	São Paulo, Brasilia, Rio de Janeiro, Natal, Fortaleza, Belo Horizonte, Florianopolis, Porto Alegre, Salvador	Guarulhos, Viracopos, Galeao, Grande Natal, Presidente Juscelino Kubitschek, Hercílio Luz, Confins, Salgado Filho, Deputado Magalhães, Pinto Martins	C
Chile	Santiago, Arica, Iquique, Antofagasta, Calama, Atacama, La Serena, Concepción, Temuco, Puerto Montt, Punta Arenas	Arturo Merino Benítez, Chacalluta, Diego Aracena, Andres Sabella, El Loa, Atacama, La Florida, Carriel Sur, La Araucania, El Tepual, Carlos Ibáñez del Campo	C
Colombia	Bogotá, Cali, Barranquilla, Santa Marta, Cartagena, Medellín and another 11 cities	El Dorado, Bonilla Aragon, Ernesto Cortíssoz, Simon Bolivar, Rafael Nuñez, Jose Maria Cordova and another 11 airports	C
Costa Rica	San Jose and Liberia	Juan Santamaria and Ernesto Cortíssoz	C
Dominican Republic	Santo Domingo, Puerto Plata, Barahona, Samana	Las Americas, Joaquín Balaguer, Gregorio Luperón, Maria Montez, Juan Bosch, Arroyo Barril	C
Ecuador	Guayaquil, Quito, Galapagos	Jose Joaquin de Olmedo, Mariscal Sucre, Isla Baltra	C
Honduras	Tegucigalpa, San Pedro Sula, Roatan y La Ceiba	Toncontin, Ramon Villeda Morales, Juan Manuel Galvez, Goloson	C
Jamaica	Kingston and Montego Bay	Norman Manley International and Sangster International	C
Mexico	Cancun, Monterrey, Guadalajara and another 31 cities	Cancun International, General Mariano Escobedo, Miguel Hidalgo y Costilla and another 31 airports	C
Peru	Lima and another 17 cities	Jorge Chávez International and another 17 regional airports	C
Puerto Rico	San Juan	Luis Muñoz Marín	C
Uruguay	Montevideo and Punta del Este	Gral. Cesáreo L. Berisso and Laguna del Sauce	C

The landscape of passenger traffic share according to airport ownership is set to change with the upcoming PPP transactions in the region, with the French government having recently passed a law to further disengage from the majority of shares in Aéroports de Paris[11] and ongoing negotiations to privatise Munich Airport, which is still owned entirely by SOEs.

PPP concessions are very popular in Latin America and the Caribbean, with the most important airports of the region already being operated by private parties under long-term concession contracts: 35 airports in Argentina (including AEP and EZE), 11 airports in Chile (including SCL), 10 airports in Brazil (including GRU, GIG, VCP and BSB), 18 airports in Peru (including LIM), 3 airports in Ecuador (including GYE and UIO), 16 airports in Colombia (including BOG) and 34 airports in Mexico (including MTY, GDL and CUN, but excluding México City), among others.

Such a widespread trend in the past 30 years has had an impact over the ownership landscape of airports in the region: while airports are still owned by states, more than two-thirds of the region's total passenger traffic (with an aggregated volume of traffic of 525m passengers in 2017) was processed at concessioned airports. This trend of airport PPP concessions is set to further continue in the upcoming years, with Brazil having recently awarded another 12 airport concessions through competitive biddings.[12]

There is still uncertainty about whether major airports in the region such as Mexico City, Panama City, San Salvador and Guatemala City will be concessioned to the private sector in the upcoming years. If this were to occur, it would imply a relevant increase in the share of PPP airports over the region's total passenger traffic.

Airports operated and owned by SOEs are still the norm in Asia-Pacific, where airports operated by public entities still represent over 80% of the region's total passenger traffic (estimated at 2,632 million passengers in 2017). One of the main reasons for this is that China's airports are all owned and operated by public entities.

Australia, Malaysia and New Zealand are the most relevant examples of airports having been privatised through sale of shares. In the cases of Australia and New Zealand, the major airports of these countries have been entirely privatised, with the state no longer holding shares or being involved in their operation. Airports with more than 50% of private sector shareholders represent 9% of the region's passenger traffic.

---

11 France's National Assembly has passed a law that provides for the removal of the obligation for the State to hold the majority of the capital of AdP, currently at 50.63%, for an amount valued at €9.5 billion.
12 In March 2019, the Brazilian Government launched a new round of twelve airport concessions across the country (on top of the ten airports that have already been concessioned), representing a total traffic volume of 19.3 million annual passengers. Airports concessioned to the private sector in this round are Cuiaba, Rondonopolis, Sinop, Alta Floresta, Macae, Vitoria, Recife, Maceio, Joao Pessoa, Aracaju, Juazeiro do Norte and Campina Grande.

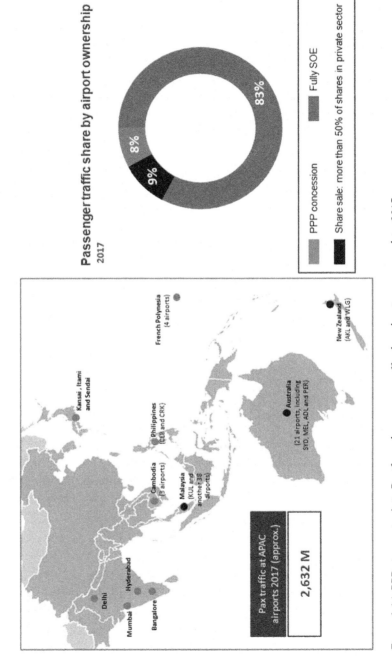

Figure 6.14 PPP airports in Asia-Pacific and passenger traffic share by airport ownership, 2017
Source: Author research, 2017 data

Table 6.6 Airport PPPs in Asia Pacific

Country	Cities	Airports	PPP Scheme*
Australia	Sydney, Melbourne, Adelaide, Perth and another 17 cities	Kingsford Smith, Tullamarine, Adelaide International, Perth International and another 17 airports	SS
Cambodia	Siem Reap, Sihanoukville and Phnom Penh	Siem Reap, Sihanoukville and Phnom Penh	C
French Polynesia	Tahiti, Bora Bora, Raiatea and Rangiroa	Tahiti Fa'a'a, Bora Bora, Uturoa and Rangiroa	C
India	New Delhi, Mumbai, Hyderabad and Bangalore	Chhatrapati Shivaji, Indira Gandhi, Rajiv Gandhi and Kempegowda	C
Japan	Kansai, Osaka and Sendai	Kokusai, Itami and Sendai	C
Malaysia	Kuala Lumpur and another 37 cities	Kuala Lumpur International, Kota Kinabalu, Penang, Kuching, Senai and another 32 airports	SS
New Zealand	Auckland and Wellington	Auckland International and Wellington International	SS
Philippines	Mactan Cebu and Clark	Clark International and Mactan Cebu International	C

Source: Author research, 2017 data

*: C = Concession; SS = Privatisation/Share Sale; MC = Management Contract

The most relevant example of PPP concessions in the region are in India, where four of the country's largest airports (New Delhi, Mumbai, Hyderabad and Bangalore) have been concessioned to private operators. The PPP agenda in this country is set to continue, with the government preparing to concession another six airports through PPP schemes (Ahmedabad, Jaipur, Lucknow, Guwahati, Thiruvananthapuram, and Mangaluru). Other relevant example of PPP concessions in the region are Mactan Cebu and Clark airports in the Philippines, Kansai and Sendai airports in Japan, Phnom Penh and Siem Reap airports in Cambodia and four airports in French Polynesia, operated by Egis.

Airports with PPP concessions represent roughly 8% of the region's passenger traffic.

Private sector involvement in airport ownership and operation in Africa and the Middle East is still incipient, with PPP concessions in place in Tunisia (two airports), Madagascar (two airports), Abidjan, Libreville, Congo (three

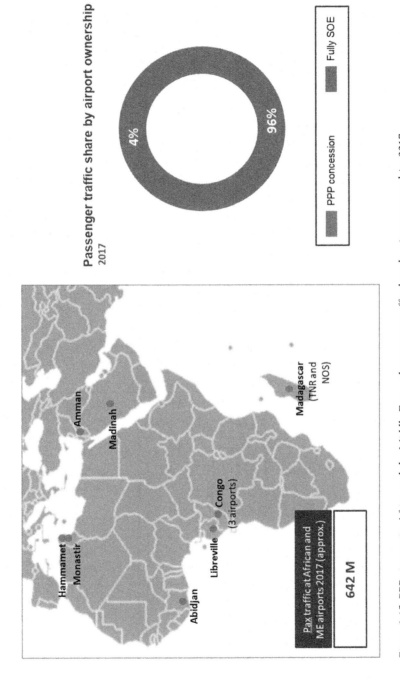

Figure 6.15 PPP airports in Africa and the Middle East and passenger traffic share by airport ownership, 2017

Source: Author research, 2017 data

Table 6.7 Airport PPPs in Africa and the Middle East

Country	Cities	Airports	PPP Scheme*
Tunisia	Hammamet and Monastir	Enfidha and Habib Bourguiba airports	C
Madagascar	Antananarivo and Nosy Be	Ivato and Fascene airports	C
Saudi Arabia	Madinah	Prince Mohammad Bin Abdulaziz Airport	C
Congo	Brazzaville, Pointe-Noire and Ollombo	Maya Maya, Agostinho Neto and Ollombo airports	C
Gabon	Libreville	Leon M'ba	C
Cote D'Ivoire	Abidjan	Port Bouet Airport	C
Jordan	Amman	Queen Alia International Airport	C

Source: Author research, 2017 data
*: C = Concession; SS = Privatisation/Share Sale; MC = Management Contract

airports), Madinah and Amman. PPP schemes only account for approximately 4% of the region's total passenger traffic, estimated at 642 million passengers in 2017.

6.5 How to select among forms of private sector participation

Decision making should focus on the main objective rather than trying to "optimise" everything:

- To generate short-term cash for a profitable airport, a sale of share or equity may be the best option
- To attract private investment and improve management, a concession or BOT is the option to consider
- To improve service and efficiency without private investment, a management contract may be more appropriate.

Only a solid feasibility study can determine if a specific objective can be achieved by a given model of private sector participation.

"Value for money" (VfM) is a measure of the net value that a government receives from a PPP project. The assessment of VfM helps the government decide whether a project should be implemented as a PPP and how much support the government should provide to that project. Assessing VfM is as much an art as a science, given the various and changing concepts of "value" that the government will want to access and apply to PPP.

> **Box 6.3 Quantitative versus qualitative VfM**
>
> The UK National Audit Office emphasis that financial appraisal and cost modelling is just one part of the overall assessment of projects, and have sought to discourage appraisers striving for disproportionate levels of accuracy. Qualitative considerations – viability, desirability, achievability – should frame the approach to quantitative assessment. The quantitative assessment should form part of the overall value for money judgement rather than be seen as a stand-alone pass/fail test. Neither the quantitative or qualitative assessment should be considered in isolation.
>
> Source: Infrastructure UK, "A new approach to public private partnerships" (December 2012)

Various approaches and models endeavour to quantify VfM, in particular through public sector comparators (see Box 6.4), cost benefit analysis and shadow models (where a financial model is developed from the bidder's perspective to test likely bidder concerns). Best practice uses such quantitative analysis as important data, but looks to a qualitative analysis to respond to all relevant parameters rather than seek measurable accuracy in assessment (see Box 6.3).

> **Box 6.4 Public Sector Comparator (PSC)**
>
> A comparison between the cost of public delivery of the project and that through PPP can provide a useful mechanism is assessing value for money. But a PSC is difficult to assemble with any accuracy. In order to assess a PSC properly, full information is needed on how the project would be implemented by the public sector, including actual cost of construction, cost of operation, cost of financing and the risk borne by the public sector (which is difficult to calculate with any accuracy).
>
> For further discussion of PSCs, see www.treasury.gov.uk.

In March 2017, the government of Bermuda entered into a contract with the Canadian Commercial Corporation, which fully subcontracted to Aecon, for a 30-year concession with an investment value of US$274 million. In order to ensure it received value for money, the government implemented an extensive VfM testing regime, including an independent fairness assessment.[13]

---

13 Again, see IATA on "Airport Ownership and Regulation" (2018b).

In some cases, VfM is used as an ex-post rationalisation of a political decision to implement a project. This can jeopardise the sustainability of the project and PPP program. A robust VfM exercise at the time of project selection and procurement can protect a project from ex-post challenges.

6.6 Selecting the right partner

The proper partner for an airport PPP will depend on the characteristics of the project and on the peculiarities of the country where the airport is located. Each type of partner will have its own priorities and character.

Airport operators may take a long view, looking to expand their global presence[14] and invest in profitable businesses. Some large airport operators have decided not to invest globally anymore, limiting their participation to technical service agreements (Munich Flughafen). Others are active operators and have increased their participation where they found profitable operations (Fraport, at Lima Jorge Chávez).

Some airport operators started as global financial investors and developed the experience by acquiring PPPs (Vinci). Other global players are currently operators of PPPs that acquired the experience by running airports, originally through a qualified international airport operator (GAP Mexico).

A construction contractor will focus more on the construction contract and a relatively short-term sell-down of equity (construction companies tend to be low margin and therefore ill-suited to carrying equity investments in PPP projects). PPPs with a great deal of construction often include a construction company within the bidding consortium, although this should not be a requirement. It will also depend on whether the construction will be concentrated at the beginning of the project (where a contractor should be engaged early, possibly as part of the consortium) or throughout the concession horizon.

Consortiums often include passive investors, who do not get involved in the operation of the airport but may have the expertise and capacity to raise financing. These financial investors may seek to sell-down their financial position in the project within the first five to ten years, once the project is in operation and is proven successful.

In some countries, no foreign investor would enter into a PPP without a local partner. Some countries require a great deal of local knowledge, contacts and political connections. In other countries with a more solid institutional and legal system, foreign investors may be able to go ahead without the need for a local partner (e.g. the Santiago de Chile Arturo Merino Benítez concession, awarded in 2016 to a consortium formed by Aeroports de Paris, Vinci from France, and Astaldi from Italy).

---

14 Never underestimate the glamour of aviation and the pride involved.

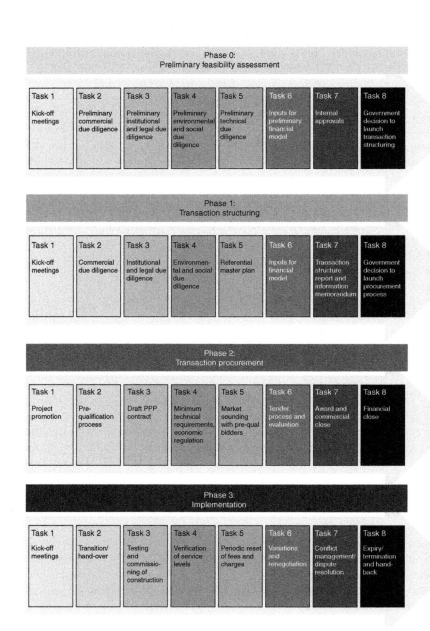

Figure 6.16 Phases of airport PPP development

Box 6.5 The importance of engaging a reputable airport operator for the airport company

Most PPPs require that the project company brings experience of airport management, a reputable airport operator with an equity stake in the project company, locked up for a significant period (often between 15 years to the whole of the concession period). Policy makers want the brand name of a reputable airport to be attached to the PPP, to ensure success. Beyond the brand name, the operator should be associated with highest standards of safety, security, reliability and level of service.

But a reputable airport operator is no guarantee of success. In some cases, despite their equity stake and contractual commitment, the reputable operator is not engaged, and does not provide skilled staff or resources. In others, the airport operators have gone well beyond their required commitment, playing a significant role in airport development and success (e.g. Fraport at Lima Jorge Chávez).

The airport operator must commit experienced managers to the project company, in a senior position. Managers in the new PPP should come from the home airport (or at least from another airport of the same operator) bringing the managerial culture, know-how and spirit of the operator. Those experts must provide time and effort to train local staff, and those local staff should spend time at the operator's main airport to learn systems, processes, procedures and culture.[15]

Many of these arrangements are achieved through a technical service agreement (TSA) which details the necessary expertise of each of the proposed managers, their experience and their proposed dedicated time. Experienced staff members serving as sporadic advisers are insufficient. Full time, relocated, foreign managers bringing their own experience from similar positions at their own airport is the best way to transfer management culture. The TSA establishes a compensation regime to create incentives for the operator to provide this management support.

In many countries a PPP airport has developed sufficient local capacity that they successfully bid to develop PPP projects in other countries managed by those local experts. Mexico's GAP (Grupo Aeropuerto del Pacifico) has been operating 12 airports in Mexico since the late 1990s, originally supported by AENA from Spain. By their own merits, GAP has qualified as an airport operator and was awarded the concession to operate Kingston Norman Manley International, in Jamaica.

---

15 At one major airport in Latin America, the operating company that awarded the concession brought a manager who was headhunted in the international airport community. "JM" proved to be an excellent manager but had never been at the home base of the airport operating company. One could ask: Why didn't the Latin American country hire JM directly in the first place?

138 *Airport PPPs around the world*

Selecting the right private partner is best achieved through the systematic selection, preparation, procurement and implementation of the project. The diagram in Figure 6.16 provides an outline of this process, and Chapters 8–11 describe this process in more detail.

7 Financing airport PPP

Always drink upstream from the herd.
Will Rogers (1879–1935), American humourist and entertainer

PPP allows governments to attract new sources of private financing and management while maintaining a public presence in ownership and strategic policy setting. These partnerships can leverage public funds and offer the advantage of contracting with well-qualified private enterprises to manage and deliver infrastructure services. International aeronautical fees are generally US dollar based, providing valuable foreign exchange revenues. Certainly these foreign exchange denominated revenues can increase the project bankability and the potential for foreign currency denominated debt from international lenders. These lenders are usually more familiar with the airport business than local banks and therefore less risk averse on issues related to traffic and operations. This explains why a Canadian teachers' pension fund might invest in an international airport in Costa Rica.

An airport PPP must be financially sustainable to attract private financing; its revenues need to be resilient and able to cover all operating expenses, including debt servicing, and provide shareholders with reasonable dividends. Lenders will be concerned to ensure that the project is able to repay interest and principal. They will have a conservative view on assumptions such as traffic forecasts, and impose specific requirements (maintenance funds, reserve fund for debt service, minimum revenue guaranteed) to provide them with additional protections, which will have financial implications. Government may have to provide minimum revenue guarantees to ensure that the airport will be able to cover all expenses during operations.

Ultimately, the cost of infrastructure has to be borne by its users or by taxpayers, current or future (aside from the limited concessionary component of foreign aid). The investments of public infrastructure firms have traditionally been financed from the public budget (through taxation), possibly with a contribution from the enterprises' retained earnings (from consumers). PPP offers alternatives to attract new sources of private financing and management while maintaining a public presence in ownership and strategic policy-setting. These partnerships can leverage public funds and offer advantages of contracting with well-qualified private enterprises to manage and deliver infrastructure services.

Generally speaking, a sovereign government will be able to obtain financing at a lower cost than the sponsors or the project company.[1] The cost-effectiveness of government financing will depend on the credit profile of the government in question (as reflected in its credit rating) and any other restrictions that apply to that government in relation to assuming new debt obligations. However, government financing is often rendered less efficient by public procurement processes; failure or unwillingness to implement incentive mechanisms to achieve greater efficiency; and failure to control changes and other risks that result in higher construction and operation costs. Private sector financing may therefore prove – in certain circumstances – less expensive, less time-consuming and more flexible to arrange or more practical than public sector financing. The private sector can provide new sources of finance (in particular where fiscal space or other constraints limit availability of government financing), impose clear efficiency incentives on

---

1 Lower interest rates obtained by a government reflect the contingent liability borne by taxpayers (See Klein's working paper on "Risk, Taxpayers and the Role of Government in Project Finance" (1996).) Thus, the risk that results in higher private finance interest rates reflects actual project risk and is subsidised by taxpayers to achieve the lower public finance interest rates. Since the private sector is best placed to manage most of the commercial risk in infrastructure projects, it is argued that private finance is the most efficient method of financing infrastructure; the inherent subsidy of public finance is more appropriately used in other areas.

the project, and can be used to invigorate local financial markets. The different types of borrowing are discussed in Section 7.1.

Government will need to consider the tenor of debt available for the project. Major infrastructure investments need long-tenor debt to reflect the long lifecycle of the asset. Where only short-term debt is available, someone will need to take the risk of refinancing. This is a difficult risk for the private sector to bear, and is unlikely to be cost-effective. Government is in a better position to bear the risk of refinancing but will need to be closely involved in the structuring of the project, structuring of the deal and negotiations with lenders in order to protect itself from this risk. Refinancing gains, on the other hand, should be shared.

The overall interest rate applicable to projects financed using corporate financing must take into consideration the minimum level of return on investment (ROI) demanded by sponsors to forego other investment opportunities. The types of financing are discussed in Section 7.2, below. The corporate entity in question will borrow funds to finance the project, but will compare the return earned from such financing against its other commercial activities, where it would invest these funds if it did not invest them in the project (the *opportunity cost* of the project). This minimum ROI (which represents the cost of corporate financing) will normally significantly exceed the cost of project financing or government financing. Corporate financing is also less project specific than project financing, and therefore may fail to implement the project-specific efficiencies and discipline generally mandated by project specific limited recourse financing.

Project financing tends to attract a higher rate of interest than government financing, since the lenders take an element of commercial risk, but lower than corporate financing where the returns needed to justify its diversion of investment funds from other opportunities. In particular, project financing offers a lower weighted cost of capital (WACC),[2] mixing cheaper limited recourse debt with more expensive private capital. It brings the benefit of:

- Highly structured risk management (including insurances)
- Fixed construction costs and time
- Fixed, or tightly controlled, operating costs; and
- Reduction of costs if the services are not delivered to specification.

As discussed, these efficiencies are embedded in the project finance structure and form an integral part of the lenders' security package. Project financing is described in more detail in Section 7.3.

Financing which appears on the balance sheet of either the host government or the project sponsor will have implications for other transactions undertaken by the government or the project sponsor in that further financing will be more difficult and more expensive to obtain. By placing the debt on the balance sheet of a special purpose vehicle in a manner that is not (or is only to a limited extent)

---

2 See Section 5.2.

consolidated onto the project sponsors' balance sheet or the government's liabilities, the debt becomes *off-balance sheet*. For this reason, the actual cost of on-balance sheet financing may be greater than perceived. Project financing may enable the government and the project sponsor to finance the project off-balance sheet and therefore avoid these costs and risks.

The implications of project financing on a government or project sponsor will depend on the accounting treatment, and therefore the accounting standards, applied. Also, it should be noted that keeping debt off-balance sheet does not reduce actual liabilities for the government and may merely disguise government liabilities, reducing the effectiveness of government debt monitoring mechanisms. As a matter of policy, the use of off-balance sheet debt should be considered carefully, and protective mechanisms should be implemented accordingly.[3]

7.1 Types of borrowing

Three of the more common sources of financing for infrastructure projects are:

1. *Government financing*: where the government borrows money and provides it to the project through on-lending, grants, subsidies or guarantees of indebtedness. The government can usually borrow money at a lower interest rate, but is constrained by its fiscal space (the amount it is able to borrow) and will have a number of worthy initiatives competing for scarce fiscal resources; the government is also generally less able to manage commercial risk efficiently. Typically, US airports borrow money to carry out expansion projects
2. *Corporate financing*: a company borrows money against its proven credit profile and ongoing business (whether or not that debt is secured against specific assets or revenues – whether the debt is "secured" or "unsecured") and invests it in the project. Utilities and state-owned enterprises often do not have the needed debt capacity, and may have a number of competing investment requirements. External investors may have sufficient debt capacity, but the size of investment required and the returns that such companies seek from their investments may result in a relatively high cost of financing and therefore can be prohibitive for the contracting authority; and
3. *Project financing*: where non- or limited recourse (these terms can be used interchangeably) loans are made directly to a special purpose vehicle. The lenders rely on the cash flow of the project for repayment of the debt; security for the debt is primarily limited to the project assets and revenue stream. The debt can therefore be off-balance sheet for the shareholders, and possibly also for the contracting authority. For example, Montego Bay, Jamaica, was developed with equity and debt from the IFC and other commercial lenders.

---

[3] Irwin, "Controlling Spending Commitments in PPPs", in *Public Investment and Public-Private Partnerships* (Schwartz, Corbacho and Funke, 2008).

The proportion used of each source of financing and the decision as to which form of financing to adopt will depend on market availability of financing and the willingness of lenders to bear certain project risks or credit risks according to their view of how the market is developing and changing and of their own internal risk management regime.

7.2 Sources of financing

A PPP project will involve financing from various sources, in some combination of equity and debt.

7.2.1 Equity contributions

Equity contributions are funds invested in the project company which comprise its share capital and other shareholder funds. Equity holds the lowest priority of the contributions, e.g. debt contributors will have the right to project assets and revenues to meet debt service obligations before the equity contributors can obtain any return or, on termination or insolvency, any repayment; and equity shareholders cannot normally receive distributions unless the company is in profit. Equity contributions bear the highest risk and therefore potentially receive the highest returns.

7.2.2 Debt contributions

Debt can be obtained from many sources, including commercial lenders, export credit agencies, bilateral or multilateral organisations, bondholders (such as institutional investors), shareholders and sometimes the host country government. Unlike equity contributions, debt contributions have the highest priority amongst the invested funds (e.g. senior debt must be serviced before most other debts are repaid). Repayment of debt is generally tied to a fixed or floating rate of interest and a program of periodic payments.

The source of debt will have an important influence on the nature of the debt provided.[4] PPP generally involves the construction of high value, long-life assets with stable revenues, and therefore seeks long-term, fixed interest debt. This debt profile fits perfectly with the asset profile of pension funds and other institutional investors.

Bond financing allows the borrower to access debt directly from individuals and institutions, rather than using commercial lenders as intermediaries.[5] The issuer (the borrower) sells the bonds to the investors. The lead manager helps the issuer to market the bonds. A trustee holds rights and acts on behalf of the investors, stopping any one investor from independently declaring a default. Bond financing generally provides lower borrowing costs and longer tenors (duration),

---

4 For further discussion of different lenders, see section 3.3.1.
5 For further discussion of documenting a bond issue, see Vinter and Price, *Project Finance: A legal guide* (third edition, 2006).

if the credit position of the bond is sufficiently strong. Rating agencies may be consulted when structuring the project to maximise the credit rating for the project. Rating agencies will assess the riskiness of the project, and assign a credit rating to the bonds which will signal to bond purchasers the attractiveness of the investment and the price they should pay.

Different types of credit enhancement can be used to increase availability and reduce the cost of debt. For example, a multilateral lending agency (MLA) like the World Bank may provide credit enhancement to bond investors. The guarantor has a superior credit rating, and provides some undertaking to investors using that superior credit rating to reduce risk for investors, thereby improving the rating for the bond and reducing the yield required, justifying the cost of the guarantee.[6]

7.2.3 Mezzanine/subordinated contributions

Located somewhere between equity and debt, mezzanine contributions are accorded lower priority than senior debt but higher priority than equity. Examples of mezzanine contributions are subordinated loans and preference shares. Subordinated loans involve a lender agreeing not to be paid until more "senior" lenders to the same borrower have been paid, whether in relation to specific project revenues or in the event of insolvency. Preference shares are equity shares, but with priority over other "common" shares when it comes to distributions. Mezzanine contributors will be compensated for the added risk they take either by receiving higher interest rates on loans than the senior debt contributors and/or by participating in the project profits or the capital gains achieved by project equity. Use of mezzanine contributions (which can also be characterised as quasi-equity) will allow the project company to maintain greater levels of debt to equity ratio in the project, although at a higher cost than senior debt. Shareholders may prefer to provide subordinated debt instead of equity to:

- Benefit from tax deductible interest payments
- Avoid withholding tax
- Avoid restrictions on some institutions not permitted to invest in equity
- Allow the project company to service its subordinated debt when it would not be permitted to make distributions; and
- Permit shareholders to obtain some security, for example to rank senior against trade creditors.

But, unlike equity holders, subordinated lenders:

- Do not share in profits
- Do not normally have voting and control rights; and
- May be subject to usury laws on the amount of interest they are allowed to charge, where equity distributions would not.

---

6 More on this at their website: www.worldbank.org/guarantees.

7.3 Project finance

One of the most common, and often most efficient, financing arrangements for PPP projects is "project financing", also known as "limited recourse" or "non-recourse" financing. Project financing normally takes the form of limited recourse lending to a specially created project vehicle which has the right to carry out the construction and operation of the project. One of the primary advantages of project financing is that it can provide off-balance sheet financing, which will not affect the credit of the shareholders or the contracting authority, and shifts some of the project risk to the lenders in exchange for which the lenders obtain a higher margin than for normal corporate lending. Project financing achieves a better/lower weighted average cost of capital than pure equity financing. It also promotes a transparent risk sharing regime, and creates incentives across different project parties, to encourage good performance and efficient risk management.

Box 7.1 Weighted average cost of capital (WACC)

This is the total return required by both debt and equity investors expressed as a real post-tax percentage on fund usage. WACC is used to measure the project company's cost of capital: the value of its equity plus the cost of its debt. WACC is calculated by multiplying the cost of each capital component, such as share capital, bonds and long-term debt, by its proportional weight and then adding these components together. Assuming that interest charged on debt is much lower than the returns sought by equity investors, increasing the amount of debt also increases equity return. This is because the total amount paid by the project company in respect of its debt and equity (measured by its WACC) will be lower compared to a project that has been fully equity financed, thereby leaving the project company with more funds for distribution and an increased return on equity (ROE). It also allows investors to spread precious equity capital over a greater number of projects (as total equity investment required for each project decreases through better leverage).

Consider the following two scenarios where $400 is invested in a project company. In scenario 1, the amount is invested fully as equity with a return of 10%. In scenario 2, the project company uses the leverage offered by a 3:1 debt to equity ratio and so the $100 equity investment is accompanied by debt of $300 at a cost of say 7%. The WACC in scenario 2 would therefore be [($100 × 10%) + ($300 × 7%)]/$400 = 0.078 or 7.8%.

7.3.1 Off-balance sheet and limited recourse financing

Project finance debt is held by the project company, which is a sufficiently minority subsidiary so as not to be consolidated onto the balance sheet of the respective shareholders. This reduces the impact of the project on the cost of the shareholder's existing debt and on the shareholder's debt capacity, releasing such debt capacity for additional investments. To a certain extent, the contracting authority can also use project finance to keep project debt and liabilities "off-balance sheet", taking up less fiscal space.[7] Fiscal space indicates the debt capacity of a sovereign entity and is a function of requirements placed on the host country by its own laws, or by the rules applied by supra- or international bodies (such as the International Monetary Fund) or by market actors (such as credit rating agencies).

Another advantage to the shareholders of project financing is the absence, or limitation, of recourse by the lenders to the shareholders. The project company is generally a limited liability special purpose project vehicle (SPV). The shareholders in a limited liability company are only at risk for the amount of equity they invested in the company. No other company liabilities attach to the shareholders unless there is some activity that would "pierce the corporate veil". The SPV is a new company with the project as its sole purpose, providing lenders with greater confidence that no other creditor outside the project compete for project revenues to pay their liabilities or for project assets in the event the company is wound-up. Therefore, the lenders' recourse will be limited primarily or entirely to the project assets. The SPV also gives the government comfort that no creditor outside the project will make financial claims against the company, or call for it to be wound-up. The extent to which some recourse is provided to shareholder assets is commonly called "sponsor support", which may include contingent equity or subordinated debt commitments to cover construction or other price overruns.

7.3.2 Bankability

Bankability encapsulates the characteristics of a project that provide lenders with confidence that the project will succeed and the lending secured sufficient to attract financing. This concept applies to all kinds of financing, but is more important and better defined when it comes to limited recourse financing, because the lenders' recourse for repayment of debt will be limited primarily to the revenue flow from the project. Due to the limited recourse and highly leveraged nature of project financing, any interruption of the project revenue stream, or additional costs not contemplated in the project financial plan, will directly threaten the ability to make debt service payments. This makes lenders extremely risk averse. The lenders will want to ensure that the risks borne by the project company are limited and properly managed and that the project involves a solid financial, economic and technical plan. Therefore, before committing

---

7 It should be noted that keeping debt off-balance sheet does not necessarily reduce actual liabilities for the government and may merely disguise government liabilities, reducing the effectiveness of government debt monitoring mechanisms. As a policy issue, the use of off-balance sheet debt should be considered carefully and protective mechanisms should be implemented accordingly.

themselves to a project, the lenders will perform an in-depth review of the viability of the project (their "due diligence"). This is known commonly as verifying the project's "bankability".

Bankability requirements will vary based on the identity of the lenders, who will have different interests and concerns and perceptions of risk. The lenders' vigilance is a key benefit of project finance, helping the contracting authority and shareholders alike to assess project viability. Clearly, an overly anxious lender can delay, complicate or even undermine a project. Equally, a lender that is not sensitive to risk e.g. where the government provides a comprehensive guarantee of the debt, will not be as concerned about due diligence and therefore the benefits of the lender's incentive to assess and monitor the project are lost. A lender's due diligence will include the following.

Economics and politics

The lenders will wish to review the effect that the local economy and the project will have on each other. Although it is the contracting authority and not the lenders who should be verifying that the project will have an overall beneficial impact on the site country and the local economy, the lenders will need to assess the net political and socio-economic benefit the project can have on the site country generally.[8] A commonly used measurement is the economic internal rate of return (EIRR[9]), which means the project's rate of return after taking into account economic costs and benefits, including monetary costs and benefits. EIRR captures the externalities (such as social and environmental benefits) not included in financial IRR[10] calculations.

The lenders will also consider political context, for example the history of expropriation of airport concessions is primarily related to change of government philosophy of PPP (e.g. Isla Margarita Porlamar, Venezuela and three large airports in Bolivia), new governments hoping to extract new concessions from the private sector (e.g. Hungary Ferihegy, Manila Ninoy Aquino, Bucharest Henri Coandă, Quito M. Sucre) leading to renegotiation or expropriation or disagreement with the tactics adopted by previous governments (e.g. Male, Maldives).

The lenders will also use this macro-level assessment to ask certain fundamental questions about the project, e.g.:

- Historical and likely future trends in fees and charges, costs, competition, demand and the nature of the demand (e.g. the financial situation of the national airline based at the airport and history of timely payments, development of other modes of transportation that could erode traffic demand)

---

8 For further discussion of this issue, see Haley, *A-Z of BOOT* (1996), p. 34.
9 EIRR is the project's internal rate of return after taking into account externalities (such as economic, social and environmental costs and benefits) not included in normal financial IRR calculations.
10 FIRR is the discount rate that equates the present value of a future stream of payments to the initial investment. See also economic IRR.

- Flexibility, sophistication, skill and depth of the labour market
- Historical and likely future trends in inflation rates, interest rates, foreign exchange cost and availability of labour, materials and services (such as water, power and telecoms)
- Condition of local infrastructure; and
- Use of land (ownership, availability of land for expansion, access to needed land within the premises – e.g. the national airline in distress holds property at the airport, social encroachments).[11]

Legal/regulatory

The lenders will want to consider the legal system (including regulation and taxation) applicable to the project in view of:

- A long-term commercial arrangement based on undertakings by the public sector, property rights, taking of security, asset management, likely tax exposure and corporate structures, and the likelihood of changes in law and taxation during the project
- Other taxes or duties that could affect traffic demand (e.g. VAT on tickets, exit duties, taxes on fees and charges)
- Whether and to what extent the legal system is accessible to the project company and the lenders, including the time and resources required to access judicial review and whether such decisions can be enforced (courts or arbitration); and
- Availability of security rights and priority given to creditors.

Business plan

In asset financing, it is the value and rate of depreciation of the underlying assets which defines the lenders' security and willingness to finance a project. In project financing, it is the viability of the project structure, the business plan and the forecast revenue stream that will convince the lenders to provide financing. Lenders will assess carefully the credit risk of different project counter-parties, including of course the project shareholders. Financial due diligence will include issues associated with financing risk such as historical information on exchange rate movements, inflation, interest rates, availability of insurance and reinsurance, and remedies available against different counter-parties for losses or damages.

The lenders will develop their financial model from the information available. This model will identify the various financial inputs and outflows of the project. By calculating project risk into the financial model, the lenders will be able to test project sensitivities, i.e. how far the project can absorb the occurrence of a given risk, for example a 10% increase in construction costs, a 10% reduction

---

11 Modified from the UN Industrial Development Organisation's *Guidelines for Infrastructure Development through Build-Operate-Transfer* (1996), p.130.

in revenues, key airline losing connecting traffic, shift in exchange rates, etc. When assessing financial viability, the lenders will use the financial model to test a number of financial ratios, in particular debt-to-equity, debt service cover, loan life cover and rates of return (some of these concepts are defined in Box 7.2).

> **Box 7.2 Financial terms**
>
> A number of financial ratios are used to test financial viability, including:
>
> - *Debt to equity ratio*: compares the amount of debt in the project against the amount of equity invested
> - *Debt service cover ratio (or "DSCR")*: measures the income of the project which is available to meet debt service (after deducting operating expenses) against the amount of debt service due in the same period. This ratio can be either backward or forward looking
> - *The loan life cover ratio (or "LLCR")*: the net present value of future project income, available to meet debt service over the maturity of the loan against the amount of debt.

Technical

The review of the project carried out by the lenders will also focus on the technical merits of the design or intended design, and the technology to be used in the project. They will want to have relatively accurate performance forecasts, including operation, maintenance and life-cycle costs, the capacity of the technology to be used and its appropriateness for the site and the type of performance required from the project. Technical due diligence will also consider administrative issues, such as the likelihood of obtaining permits and approvals using the technology in question, and the reasonableness of the construction schedule and cost. Environmental concerns, including noise pollution, can have a significant impact, see for example the environmental concerns blocking the new airport in Mexico, community engagement for the proposed Heathrow expansion, and the night curfew imposed on Frankfurt.

7.4 Refinancing

After completion of construction, once construction risk in the project has been significantly reduced, the project company will look to refinance project debt at a lower cost and on better terms, given the lower risk premium. In developed economies, the capital markets are often used as a refinancing tool after completion of the project, since bondholders prefer not to bear project completion risk, but are often able to provide fixed rates at a longer tenor and lower margin than commercial banks. Refinancing can be very challenging, in particular for lesser developed financial markets, but can significantly increase equity return, with the excess debt margin released and the resultant leverage effect, where the project performs well and where credit markets are sufficiently buoyant. Refinancing also benefits the economy, by releasing debt capacity of the original financiers for other productive uses, including new PPP projects. While wanting to incentivise the project company to pursue improved financial engineering, in particular through refinancing, the contracting authority will want to share in the project company's refinancing gains, often in the form of a 50/50 split. The contracting authority may also want the right to require refinancing in certain circumstances.

7.5 Financial intermediation

The government may wish to use its support to help mobilise private financing (in particular from local financial markets), where that financing would not otherwise be available for infrastructure projects. The government may want to mobilise local financial capacity for infrastructure investment, to mitigate foreign exchange risk (where debt is denominated in a currency different than revenues), to replace retreating or expensive foreign investment (for example, in the event of a financial crisis) and/or to provide new opportunities in local financial markets. But local financial markets may not have the experience, or risk management functions, needed to lend to some sub-sovereign entities or to private companies on a limited recourse basis.

To overcome these constraints, the government may want to consider the intermediation of debt from commercial financial markets, creating an intermediary sufficiently skilled and resourced to mitigate the risks that the financial markets associate with lending to infrastructure projects. To achieve this, the government may want to use a separate mechanism (the "intermediary") to support such activities without creating undue risk for the local financial market, for example, by:

- Using the intermediary's good credit rating to borrow from the private debt market (e.g. providing a vehicle for institutional investors who could not invest directly in projects) and then lend these funds to individual entities or projects as local currency private financing of the right tenor, terms and price for the development of creditworthy, strategic infrastructure projects
- Providing financial products and services to enhance the credit of the project and thereby mobilise additional private financing, for example by providing the riskiest tranche of debt, providing specialist expertise needed to act as lead financier on complex or structured lending, syndication, credit enhancement, and specialist advisory functions; and/or
- Providing support to finance or reduce the cost or improve the terms of private finance for key utilities. These entities may need first to learn gradually the ways of the private financial markets, and the financial markets may need to get comfortable with lending to infrastructure operators. This

mechanism can help slowly graduate such sub-national entities or state-owned enterprises from reliance on public finance to interaction with the private financial markets.

Current best practice indicates that such intermediaries should be private financial institutions with commercially oriented private sector governance. Intermediaries meant to create space in an existing financial market must have commercial incentives aligned to this goal, with appropriately skilled and experienced staff, and a credit position sufficiently strong to mobilise financing from the market. Existing private financial institutions with appropriate skills and capacity can help to perform this function. However, private entities often suffer from conflicts of interest (e.g. holding positions in the market such that their interests are not aligned with the role of intermediary) or would be constrained from taking positions in the market due to its role as intermediary (crowding out vital market capacity). The government may therefore want to create a new private entity to play this role, despite the cost and time required.

8 Airport PPP preparation

Too swift arrives as tardy as too slow.
Shakespeare, from *Romeo and Juliet* (1597) Act II, scene 6, line 14

An airport PPP project is born of much thought and preparation, as set out below and displayed graphically in the diagrams in this chapter.

- The government will set its strategy and allocate budget for investment
- The contracting authority will identify investments best delivered through partnership with the private sector. This initial selection process is based on

little information, but must filter out those projects that are inappropriate for PPP to avoid wasting time and money
- The chosen investment needs to be tested, through a pre-feasibility and possibly a feasibility study. The government will use these studies as the basis for various government approvals and allocation of government support, including budget allocations and guarantees, if needed
- Once tested and approved, the project basic design, structure, risk allocation and agreement are developed and the bid process launched
- A preferred bidder is chosen, financiers engaged, and financial close achieved
- The project is then implemented.

This process is expensive and time consuming, but can achieve the following objectives:

- *Attracting a larger number of high-quality bidders*: well developed projects bid through fair and transparent bid processes, brought to market at a time when there are not too many competing opportunities,[1] attract more competition from high quality firms. This is particularly true in countries perceived to be higher risk (for example which lack rule of law[2]) or difficult economic times, where investors become more selective, resulting in a flight to quality
- *Increasing access to financing*: lenders are attracted to well-run bid processes and the well-prepared bids they tend to engender. Competition amongst lenders results in lower cost of financing and well-prepared financing arrangements
- *Mobilising more investment*: balancing public and private financing of the project contributes to project bankability and value for money for the government. A strategy focused solely on maximising private investment may increase the project risk profile and deter potential bidders. A more efficient approach is to understand what can be mobilised from the private sector and what type of government support is needed – maximising value for the government
- *Improving the likelihood of success of the project*: successful PPP projects identify and address risks in advance, which provides comfort to all stakeholders. Well-prepared projects are more likely to have identified risks, addressed those risks, or created the dynamics needed to address risks as they arise. Good preparation anticipates issues, addresses loose ends, reduces conflict and resolves disputes.

Poor project preparation can often be traced to tensions within government between those who seek rapid results and those willing to invest in more thorough preparation. This tension is understandable given the time requirements

---

1 Resources for the usual bidders are limited and with a few competing processes at the same time they might have to opt.
2 Countries that have failed to protect investors in the past, including expropriation, may find it more difficult to attract high quality bidders.

for preparation and bidding. In reality, poorly prepared projects do not result in or speed progress and often do not achieve financial closure, but as the delays lengthen, pressure mounts to find shortcuts in the project preparation process.

Although it may be appealing to try to shorten the preparation and bidding schedule, or even to select a specific operator without competition, international experience has shown that such strategies often result in higher costs, delays due to extensive negotiations, and a perception of lack of transparency from the public, which makes the project vulnerable to legal and political challenge or even cancellation (in particular with a change in government or leadership).

Good preparation helps validate the project by testing assumptions and identifying challenges – allowing government to decide whether to go forward. This process starts with project identification/selection to ensure that only viable projects with a likelihood of success are prepared for PPP. During project identification and selection, a clear rationale should be defined for the project, with clear strategic objectives. For example, government strategy for PPP may seek to achieve a number of different objectives for airport operations, including investments, improved levels of service, increased efficiency, increased revenue and reduced government subsidy.

Once a project is selected, a pre-feasibility study tests project fundamentals. A pre-feasibility study can be completed in about three months, assuming consultants are already selected and mobilised. Government will in addition develop an airport conceptual masterplan, to establish the basic expectations with respect to airport development. A conceptual masterplan usually requires three to four months to be completed and is generally undertaken by an experienced airport planning consulting firm.

The approved pre-feasibility study is then developed into a full feasibility study, to test the project more completely, verifying political, technical and commercial dynamics and selecting the optimal project design and structure. Governments can easily focus too early on a preferred option. Starting with a narrow set of options or a pre-determined approach denies the government the opportunity to develop a solution specific to their needs, and to innovate.[3]

During project preparation, government should develop a robust business case to determine the preferred solution, based on evidence and stakeholder and market consultation. Transparent and competitive procurement process is essential. Unsolicited proposals and sole-source transactions should be handled with special care. The evaluation process for proposals should be objective, with input from independent parties. Effective competition is improved where government is very clear on its priorities based on a thorough business case.

It usually takes six to nine months (depending on how much information from the master plan needs to be reassessed) to complete a feasibility study, which will assess:

---

3 See here the UK HM Treasury's "Green Book" (2018).

- *Technical viability*: current and expected traffic, airport development options and investment requirements, construction costs and time, architectural requirements, performance requirements, and so forth
- *Financial viability*: revenue forecasts and airport financials, scope of the PPP, regulatory constraints, bankability (considering the level of investment requirements)
- *Commercial viability*: market appetite (identity of potential investors), risk allocation plan, source, cost and terms of financing, among others
- *Legal viability*: legal framework, regulatory framework, land/property rights, jurisdiction of authorities, etc.

Based on the feasibility study, government will decide whether to start the transaction process and hire transaction advisors. A series of government internal approvals (e.g. Cabinet approvals) is usually required before a formal green light is given. This process can take anything from a week to months, and should not be underestimated.

The bidding process tests the work done during the feasibility/preparation phase. The complexity of the transaction depends on the type of contract. A management contract does not require the same time for bidders to prepare proposals as do concessions, where bidders must take into consideration not only technical but also financial and legal considerations. This increases complexity, cost and the time needed to prepare a proposal. Decision makers should plan at least six months for the bidding process for management contracts and six to twelvemonths for a concession.

Government must commit significant resources and capacity after financial close to ensure proper implementation. An implementation plan is essential to realise anticipated benefits. Continuity between the transaction team and implementation team can help effectiveness.

The elements of good project preparation are mapped out in Figures 8.1 to 8.3. Figure 8.1 shows all four phases: the two phases of project preparation: pre-feasibility and feasibility, followed by procurement and implementation. Phase 0 and 1 are then mapped out in more detail in Figures 8.2 and 8.3.

8.1 Project selection

The planning function needs to help allocate projects most effectively between public and private solutions. Equally, those responsible for investment decisions, prioritising some projects for public investments and selecting those appropriate for PPP, must have the knowledge and data needed to make those decisions.

Not all airport projects are appropriate for PPP. Governments should consider the total investment needed, the total funding available, and what part of its investment needs to develop through PPP. Governments will have many investment projects, but only a limited number of them will have potential for implementation as PPPs. The process of selecting which projects to implement as

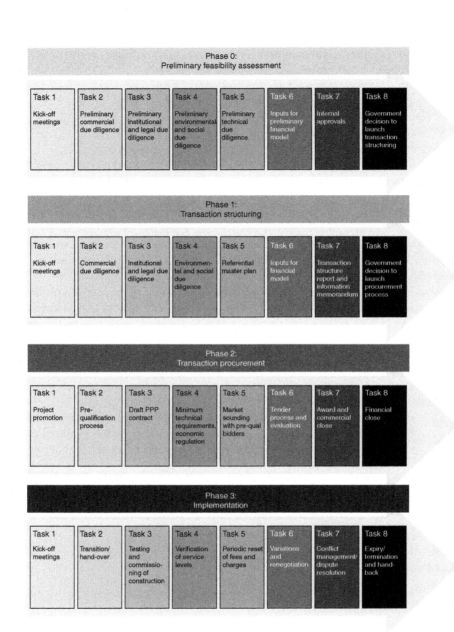

Figure 8.1 The phases of developing a PPP airport

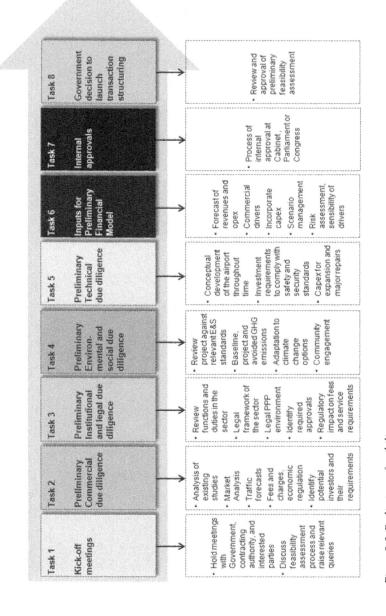

Figure 8.2 Preliminary feasibility assessment

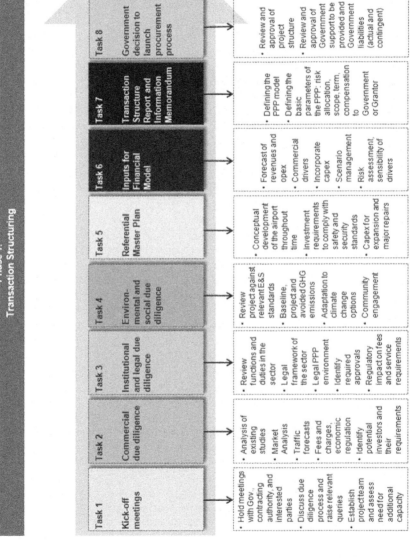

Figure 8.3 Transaction structuring

PPP (as compared to public financing or purely private) is complex, requiring a preliminary analysis of each project based on sufficient data.

Governments need to focus on those airport projects that have the greatest potential to be viable PPP projects. Given limited time and resources, selecting the wrong projects for PPP can undermine the entire program and waste government resources.

Governments should avoid the temptation to require perfect projects. This early selection process involves a limited amount of information and is therefore likely to be imprecise, but is a critical part of the decision process.

When selecting projects, and formulating a pipeline strategy, governments should note that the PPP program may need to be phased, in particular where a number of large projects are contemplated. Governments often want to implement PPP in the largest airports immediately. Investors will encourage this. The contracting authority will generally have difficulty implementing multiple projects in parallel, given the staffing, skills and financial requirements (see Box 8.1). Also, the debt markets have limited capacity and appetite. Phasing will allow the market to absorb demand without increasing costs.

8.2 Project team

A good airport PPP relies on a solid project team (supported by reputable transaction advisers) formed of financial, legal and technical professionals with demonstrated experience in successful PPP airport transactions (again, see Box 8.1). Even in the case of a less complex management contract, the government should have experienced/reputable advisers to guide it through the process. These advisers are expensive, but bring international expertise and their reputation to the project, which gives comfort to investors that the project will be well prepared. Sufficient funds should be allocated to have advisers involved until financial close. One innovative approach tested by the City of St Petersburg for the Pulkovo Airport was to hire an additional adviser (in this case the World Bank Group, though many others can provide similar services) close to the project team to focus on strategic issues, review the transaction advisers' output, and facilitate the interface between the project team and the transaction advisers.

Box 8.1 Airport PPP transaction advisers

The transaction advisers are usually separated into three teams of professionals:

1. Financial and structuring
2. Technical
3. Legal.

> Financial analysis and PPP structuring is generally performed by a separate financial adviser (typically an investment bank or an IFI). The financial analysis could be included within the scope of the technical adviser, if the firm has the relevant skills and experience. The financial adviser is often the lead adviser, and is responsible to guide the client throughout the process.
>
> Technical advisory services are provided by aviation consulting firms, comprising airport planners, engineers, and air transport economists. The technical adviser is hired early in the process, to review existing material, carry out technical due diligence, project airport throughput and define the investment requirements of the airport throughout the PPP term.
>
> The legal adviser is normally engaged at the due diligence stage, to review the legal framework, identify all legal risks of the project, and produce prequalification, contractual and tender documents for the implementation of the PPP.
>
> It is imperative that, even when high profile consulting and law firms are hired, there is a solid group of professionals with hands-on experience in airport PPPs. Each of the advisers should present a team of individual professionals with proven experience, and the dedication of each expert to a number of days and availability to develop the project. Each expert, whether engineer, planner, economist, financial analyst or lawyer, should have relevant experience in similar projects, have a minimum number of years of experience as a professional, and indicate a sufficient dedication of time to the project in days (specifying where on and off site). This should be a critical factor for selection of each adviser.
>
> Big firms that do not present a solid and specific team with proven records should be avoided. The success of project preparation will be experience, not just brand names.

A technical team will be formed to support, prepare and implement each project. The contracting authority's project team will be responsible for overseeing the preparation of the PPP project and implementing of the PPP procurement process through the final negotiation, signing of the PPP agreement, design, construction and operation of the project.

The project team is responsible for providing the key management, oversight, and decision-making functions for the PPP on behalf of the contracting authority. It will need to address key issues in PPP, such as those with the following aspects: technical, legal, financial, commercial, environmental, social, stakeholder management, regulatory, etc.

The project team will need appropriate skills and experienced, financial, legal, technical, insurance and other advisers. Each of these advisers will be subject to different agendas and incentives which will influence the nature of their advice and the ease with which the government will be able to manage their involvement. The project team (possibly through the PPP institutions) should have access to expertise in managing such advisers. To provide capacity to the project teams, the UK arranged for secondment of staff from commercial banks and law firms with expertise in project finance into the PPP unit. South Africa and Egypt initially hired long-term expert consultants who had experience in successful PPP programs to work in their PPP unit to improve access to global best practices. Many of these skills can be bought in through short term contracts and transaction advisers, though the project team will need the capacity to manage those advisers and the underlying issues to be resolved.

When establishing a project team, contracting authorities should provide it with clear authority and resources to make timely decisions on behalf of the public authority. The team will change in role, mandate, capacity, staffing and funding requirements as the project develops and moves from phase to phase of the PPP cycle.

8.3 Pre-feasibility

A pre-feasibility study (also known as an outline business case or preliminary viability study) tests the fundamentals of the project, based on a preliminary technical survey identifying key constraints and assessing the basic technical and financial project fundamentals such as concept design and possible forms of implementation, revenue and financing and site verification (in the case of a greenfield airport or major expansion). A first-level financial model will be developed at this stage, to test the viability of the project and the potential appetite of investors. As part of the pre-feasibility study, the contracting authority makes a preliminary assessment of value for money,[4] which tests the value provided by PPP.

A pre-feasibility study can be prepared quickly but usually take at least two to three months. This study provides the government with a first assessment of the project and will be sufficient to inform a "go or no-go" decision.

The pre-feasibility study will involve a preliminary assessment of three basic elements:

1. *Demand forecast*: a preliminary traffic forecast should be carried out for passengers, aircraft movements and cargo, with two different objectives:
    a. To assess airport financial performance, estimating the projected revenues and operating expenditures ("opex"), as a function of the total annual throughput. While opex typically depends on the size of airport facilities, some cost items may be influenced by traffic throughput (e.g. service contracts on security and cleaning)
    b. To assess the required capacity needed for the airport in terms of airside and landside. The capacity of the airside (runways, taxiways and aprons) will be assessed against its peak hour demand. The capacity of

---

4 For further discussion of value for money, see Section 2.11.

the terminal building will be measured by the traffic throughput during the "design busy hour" according to a target level of service.[5] Based on the capacity required on a yearly basis, architects and airport planners will define the required level of investment (capital expenditures, or "capex")

The demand forecast will need to start by a market analysis, assessing the existing and future trends with respect to air transport demand.

2. *Investment requirements* ("capex"): a team of engineers, architects and airport planners will define capital expenditures required during the entire horizon of the PPP, based on the following (see also Section 3.5.2 for a more detailed discussion of capex):
   a. Compliance with the standards and recommended practices (SARPs) defined by the International Civil Aviation Organization (ICAO) in its Annex 14 ("Aerodromes"), Annex 17 ("Security") and any relevant local regulations with respect to airport design and operations[6]
   b. Response to capacity needs, on a yearly basis, to accommodate the existing and future traffic demand, in terms of aircraft movements (airside) and passenger throughput (landside).
3. *Financial analysis*: using the previous two elements, the projected traffic throughput and the projected capital expenditures, the financial team will elaborate a full financial projection considering all airport activities and future investments, throughout the entire period of the PPP. The financial analysis will include a review of the fees and charges structure for aeronautical services. The financial analysis will identify any potential upside or shortfall in aeronautical revenues, given forecast operating costs. The analysis will also include all revenues and costs from all other non-aeronautical activities such as purely commercial activities as well as from ancillary activities such as aeronautical-related commercial activities (e.g. ground handling, in-flight catering, fuelling services, rental of spaces to airlines).

Governments can make a rapid preliminary assessment of the need for financial support with minimum information. Preliminary assessment of capital expenditure, operating costs, and traffic could be available from a previous master plan (that includes required investments) and updated traffic forecasts. Using a minimum of assumptions, it is possible to determine whether the airport is profitable enough for a concession, a management contract or some other PPP scheme. An important consideration for concessions is the availability of non-recourse financing, and any financial support needed from government to fill any financing gap. At this stage, governments should test options for providing financial

---

5 For further discussion of airport design, please see Chapter 4.
6 The reasoning for verification of compliance with SARPs derives from the fact that most countries are signatory members of ICAO and therefore those SARPs are embedded within the local legislation.

support, including balancing upfront capital contributions with operating subsidies, guarantees, etc.[7]

Once a preliminary decision to undertake the project through private investment has been made, a feasibility study is undertaken to identify key project issues and constraints.

Box 8.2 Bundling

One of the challenges of any national airport program is how to provide air transport links into lightly populated or low-traffic areas of the country, when the relevant airports do not make enough money to cover their costs, let alone pay for expansion and the kind of profits PPP might require, while still availing the airport of the skills, efficiencies and dynamism of a privately managed airport. Very few PPPs have been tried in individual small airports, given the challenges associated with high transaction costs. It is hard enough to use PPP for larger airports. But where they have been tried, the usual approach is a management contract, where most or all capital expenditure is provided by the government, with operation and maintenance provided by the private sector.

The operation of airports bundled under one single company could represent economies of scale for the investor, particularly in airport administration (accounting and finance) and use of central services, such as project development, engineering, procurement and maintenance. Another advantage of bundling is that it may allow the diversification of risk, when airports with different patterns of traffic can be put together. Such is the case of the twelve airports bundled as the Pacific Airports Group (Grupo de Aeropuertos del Pacífico – GAP), composed of airports where traffic is tourism driven (such as Los Cabos and Puerto Vallarta, representing high- and middle-end tourism), and airports more focused on business and VFR traffic (such as Guadalajara and Tijuana).

Bundling is also advantageous if the airport system depends on one (or a few) large airports. The appeal of high revenue airports may be used to attract investors to less appealing ones, by cross-subsidising. This approach transfers the responsibility of cross-subsidisation to the private sector, making the investor responsible for using the proceeds of the good airports to support the less profitable ones.

The main problem with this approach is the change in balance of power between the government and the private sector, particularly in weak regulatory environments. Too many airports in the hands of one single operator may facilitate regulatory capture (putting too much power in one investor). It may represent a much higher cost of rescuing the PPP (one big project,

---

[7] See generally, Jeffrey Delmon's *Public-Private Partnership Projects in Infrastructure* (2011).

if it fails, means one big failure). Dividing a large number of airports into several contracts diversifies the risk; if one contract has to be terminated, the airports can be redistributed among the other operators. Bundling also deprives the government of internal comparisons and competition between airports (pitting operators against each other to demonstrate greatest management efficiency).*

A one-by-one strategy takes much longer but allows governments to gain experience from each transaction and translate those lessons to develop in-house capacity and improve future projects. The Concessions Unit within the Chilean Ministry of Public Works (MOP) acquired significant experience through successive airport concessions. The one-by-one approach may also allow governments to minimise risk of mistakes by avoiding starting with the largest and most attractive airports, so they can build experience before preparing the largest projects. A successful first airport PPP project also reduces the investor's perceived risk for any subsequent deal, resulting in a better value for the government and airport users.

Bundling can also be done in groups, resulting in a hybrid of the two previously mentioned schemes. The Mexican Secretary of Communications and Transport (SCT) bundled airports in groups, each group done subsequently, with each new process implementing lessons learned in the previous processes.

* See Formulation of Policies for the Airport PPP Program of the Philippines – Department of Transport and Communications. Philippines. Andres Ricover, November 2015.

8.4 Feasibility

The feasibility study (also known as a full business case) elaborates and expands the pre-feasibility study, provides more detail, uses more rigorous analytics (including a far more detailed traffic forecast and a more thorough analysis of investments and financials) and serves as input to the bidding documents and contract to be signed with the private company/consortium. The feasibility study allows key issues to be addressed early, resulting in a more robust Information Memorandum and bid documents, which provide potential investors with an understanding of project risks, increases their appetite and stimulates competition, and most importantly, provides the government with a more accurate assessment of the value of the asset and what should be expected from the transaction.

A well-prepared feasibility study saves time and money by helping to identify areas requiring attention and suggest options that may be unavailable later; for example, improved commercial activities and the service level that can be achieved through a management contract without seeking immediate private investment (see Chapter 6 for discussion of different commercial approaches to airport PPP). This approach is intended to achieve improvement in service to airlines and passengers in the short term, while the more complex and time-consuming concession with its medium to long term benefits would be prepared in parallel.

Feasibility studies are based on the relevant airport master plan, and often will be used to update traffic forecasts and other key data of the master plan to reflect the most accurate and recent information.

Box 8.3 Factors influencing revenue potential of an airport

Factors under government control

1. Fees and charges for aeronautical services (e.g. increase existing and introduce new charges)
2. Traffic build-up, when influenced by government policies (e.g. liberalisation of aviation policies, increase in airline competition, achievement of greater connectivity and reduction of air fares, policies on receptive tourism, policies affecting local residents for outbound travel, promotion of events)
3. Facilitation of monopolistic services with elasticity of user demand (e.g. fuelling costs at the airport).

Factors outside government control

1. Traffic demand, particularly inbound foreign tourism (irrespective of local tourism policies)
2. Commercial development derived from airport traffic as well as from the local community and airport employees
3. Passenger disposable income (mostly foreigners) for consumption of commercial services
4. Environmental, political or social issues reduce inbound tourist arrivals volume
5. Better conditions for origin and destination (O&D) and connecting (implying same or different airlines) passengers from competing airports
6. Traffic volume increase through the successful development of the airport as a hub by a local airline, when the economic conditions exist for the airline (O&D mass, unattended neighbouring markets, geographical advantages)
7. Traffic volume declines from migration or distress of airlines operating hubs at the airport.

The feasibility study includes:

Institutional drivers

- Contracting authority and sector strategic objectives – whether the project is in national or local plans, meets critical public needs and has political support, including from key stakeholders, for example airlines, consumer groups, local government, etc.
- Capability of contracting authority and the project team to effectively manage the project, for example whether there is sufficient budget allocation for the project.

Economic valuation

- Fiscal affordability and consumer/end-user affordability
- Public/government/contracting authority benefits, for example, taxes, customs, duties and excise levies, employment generation, regional development, improvements in quality of life, attracting private investment (in particular foreign direct investment), economic growth, revenue share
- Government costs such as likely environmental impact, impact on the site and the community and available mitigation measures.

Financial analysis

- Revenue estimates, source of revenues (e.g. fees and charges collection), demand forecasts, fees and charges profile, credit risk of key counterparts (e.g. national flag carriers), willingness to pay, elasticity of revenues with demand/fees charged
- Cost estimates for construction, operation and financing, including inflation risk, interest rate risk, foreign exchange risk, refinancing risk (e.g. where debt/tenor is insufficient)
- Public money support approved and sufficient (amount and terms)
- Base financial model (showing return on equity, return on investment, net present value, financial internal rate of return, and debt service cover ratio, with assumptions on debt:equity, debt currencies, debt tenor inflation rate, discount rate, depreciation, interest rate, foreign exchange rate and tax risk and summary of results), including sensitivity analyses for cost increases and revenue reductions.[8]

---

8 Sensitivity analysis – determining the resilience of the financial model to changes in assumptions and risk components over the PPP term, assessing the impact of these risks; assessing the likelihood of these risks arising; calculating the value of risk (and ranges of possible outcomes); allocating risks to the party best able to manage them, and identifying strategies for mitigating risk.

Technical analysis

- Full demand assessment, with elasticity of demand against fees and charges and sensitivities for possible changes in circumstances
- Selection of process and technology, process description, engineering, layout and basic if not more detailed design, technical options, construction methods, project construction schedule, costs, time, likelihood of failure and interface with other technologies
- Performance/output specifications and whether it meets the needs and requirements of the government
- Schedule of approvals, processes, regulatory matters, licensing and permitting regime – risk of renewal, withdrawal, change in standards – in particular environmental (including equator principles[9]) and social requirements, potential blockages and critical path issues
- Access to land and process for acquisition/compensation/resettlement, with assessment of subsurface risk, archaeological remains, man-made obstacles, zoning and planning, utility supplies, nature of existing facilities, and interconnection with other facilities.

Legal assessment

- Key legal compliance challenges, including approvals, procurement, regulations, environmental and creating security rights
- Legal authority to apply PPP approach and enter into project agreements (*vires* assessment)[10]
- Key terms for all contracts and documents, including tender documents, project contracts, and project information brief
- Access to justice, including enforceability of arbitral awards, and ability of private parties to challenge government actions in court.

Comprehensive risk matrix

- For all project risks, identify the party which would be negatively affected in the event of the risk materialising, how much they would be affected, the likelihood of the risk, how those risks could be managed or mitigated, the cost of mitigation, who is incentivised to mitigate and how the risks should be allocated.

The feasibility study is intended to demonstrate "value for money" (VfM), a measure of the net value that a government receives from a PPP project. The assessment of VfM helps the government decide whether a project should be implemented as a PPP and how much support the government should provide to

---

9 See here www.equatorprinciples.org.
10 See Section 2.13.

that project. Assessing VfM is as much an art as a science, given the various and changing concepts of "value" that the government will want to access through PPP. The very definition of VfM can be adjusted to respond to changes in government priorities and requirements over time.

The process of assessing value for money is iterative. From the earliest project selection processes, government should use value for money as its standard. This assessment will gain in detail and sophistication throughout the project cycle, as more information is gathered from pre-feasibility studies, feasibility studies, procurement and implementation.

Various approaches and models endeavour to quantify VfM, in particular through public sector comparators, cost benefit analysis and shadow models (where a financial model is developed from the bidder's perspective to test likely bidder concerns). Best practice uses such quantitative analysis as important data, but will give specific consideration to a qualitative analysis to respond to all relevant parameters rather than seek measurable accuracy in assessment.

Once the government has identified the transaction objective and the financial feasibility of the business, the design of the transaction can be finalised, which includes selecting the:

- PPP model
- Scope of services to be transferred (complete airport, only landside, only airside, complete or partial airport plot, specific services, additional activities, etc.)
- Level of risk sharing between public and private sector, which will condition which model to select
- Duration (if fixed or variable, and factors influencing the term)
- Control mechanism (including governance, monitoring and regulation)
- Implementation process (bundling or separated airports, all together or step-by-step, phasing of investment, etc.).

As in most such exercises, a balance needs to be found between the time and expense of the "perfect" feasibility study, and a feasibility study that addresses enough to meet market, contracting authority and government approval requirements.

Failure to implement the different stages of project preparation properly, with sufficient time, funding and expert advice, has doomed many PPP projects and programs; this preparation process should not be curtailed. As in most such exercises, a balance needs to be found between the time and expense of the "perfect" feasibility study, and a feasibility study that addresses enough to meet market, contracting authority and government approval requirements.

Interested parties, such as potential investors, funders or contractors, may offer to develop, or may produce of their own accord, "feasibility studies". Governments need to be cautious. Even with the best of intentions, such studies will be biased towards the interests and context of the proponent. Governments will need their own, independent study to ensure feasibility is properly tested, key choices are well founded and the government has all critical information needed to negotiate with eventual investors and funders.

Having the project approved as a PPP further to a feasibility study ensures political buy-in to the process before the government and potential bidders start investing further in project development. Following the feasibility study, and associated approvals, the contracting authority is ready to commence the tender process.

8.5 Consultation

Stakeholder engagement is critical to ensure their inputs are included in the design. A long list of project options should be considered, without limitation based on identified assumptions. Community engagement is key to understand local context, community vulnerability, employee context and to obtain essential information for project design. Shortlisting of project options should only occur after a full review of context and consultation with stakeholders in the market and the local community. This consultation process can also help manage conflict and avoid social stress.

Airline consultation is an important part of this process. Engagement and sharing of proposed plans for development have to be tested with the main users for operational and financial considerations. Any possible impact on fees and charges will have to be consulted with airlines, to anticipate the market reaction and the plausibility of its implementation. Other representatives of the relevant market drivers, such as the tourism industry (e.g. tour operators, hotel associations), industrial and business community and other sectors of the society should be engaged at this stage of the process.

IATA recommends user consultation at the infrastructure development process[11] ensuring that the investment program:

- Is consistent with accepted fees and charges
- Establishes a mechanism for consultation with airlines on relevant user issues
- Considers airline functional requirements
- Shares the airport operating expenses transparently with the users.

8.6 Market sounding

Project promotion raises awareness among potential bidders and provides market feedback on concerns to be addressed. Project promotion will include (i) general marketing of the project in the press and through government's trade promotion bodies, (ii) targeted marketing "teasers" sent to key investors and presentations

---

11 See the IATA's guidelines on "Airport Infrastructure Investment: Best Practice Consultation", available on their website as a PDF.

at specialised trade events and (iii) articles in key industry publications and newsletters. Promotion should include all key market players – e.g. operators, construction companies and lenders.

Promotion of an airport project is an ongoing, iterative process. To some extent the industry will be aware of the government's need for investment and desire for improved management, but the government should make a clear statement to the market of its intention to implement a PPP solution, and the nature of its intended approach.

Project promotion starts with information gathered, as discussed in Section 8.4, above. This information can be collected and shared through promotional presentations (often called "road shows"), at conferences, on the web, and in print. These elements of the promotional process, especially in the early days of project preparation, can be used to "road test" the project, testing the approach with potential investors to refine and expand market appetite for the project. Promotional presentations can be located in the city where the airport is located, in the capital city, in key foreign cities closer to key airport operators/investors such as London, Paris, Mumbai, Istanbul, Singapore and Frankfurt, or may be focused on individual investors, at their headquarters. Industry journals and newsletters are important for the promotion process, with the on-line version of key journals reaching most if not all key investors; titles include *Air Transport World*, *Airports International*, *Airwise News*, and *Passenger Terminal World*. Journals focused on PPP include *Project Finance International*, *Project Finance Magazine* and *Infrastructure Investor*. In addition, business journals such as the *Financial Times*, *The Economist* and the *Wall Street Journal* are often good options to reach the investment community.

Promotion documentation may include a "teaser" for potential investors, with a detailed description of the project and the key terms of project documentation (see Annex 1 to this chapter for an outline of the contents of an information memorandum or "info memo"). An information memorandum is a promotional document that describes the asset, including technical description, commercial structure, demand profile, financial fundamentals, performance requirements, key risks, and how risk will be allocated.

The information memorandum will contain an indicative (non-binding) traffic forecast with some basic assumptions. In terms of financials, it will contain the current set of fees and charges and it may contain some historic financial performance of the airport. However, it must never include any kind of financial projections; delivering financial projections could be perceived as a commitment from the grantor about future financial performance and might place the bidder in an advantageous position in any negotiation.

The promotion process needs to be transparent. Any suggestion that a specific investor is preferred can alienate investors and reduce market appetite. The process can be used to evidence the level playing field that the government plans to create for bidders, which increases competition for the project and generally results in better bids and prices.

8.7 Approval regime

Approvals help raise key questions and issues during preparation of the project. They are important for quality control but also for buy-in from different agencies, achieving greater ownership and certainty for investors. But these layers of different agencies with approval rights can complicate the process. To the extent possible, these approval requirements should be streamlined, to facilitate efficient application, reduce the cost of approvals, and fast-track the investment process.

The following describes a few of the key parties that often have approval rights and the points in the project process at which such approvals are usually required.

- *Sector line ministry* will be responsible for sector strategy, economic impact (e.g. setting fees and charges, cross-subsidisation amongst consumers and cost allowances), environmental impact, consumer protection, prioritisation and risk management, in particular during project selection, feasibility verification (based on the feasibility study), before issue of bid documents, award/financial close, and renegotiation
- *Ministry of finance/budget/planning* will be responsible for allocation of government support and fiscal risk management in particular during feasibility verification (based on the feasibility study), before issue of bid documents, award/financial close, and renegotiation
- *Technical regulator* will be responsible for setting all the technical requirements to comply with safety and security standards and the oversight body throughout the PPP term
- *PPP unit or agency* may provide support to the government's PPP program, including support for project selection, prioritisation of selected projects, preparation of projects, allocation of government support and monitoring and evaluation of projects during implementation
- *Procurement agency* is often responsible for monitoring the use of transparent, competitive procurement
- *Environmental agency, land agency, central bank, attorney general, etc.* (there may be a variety of agencies with regulatory authority over specific issues).

Annex: Structure of an information memorandum

1. *Executive summary*
    1.1 Overview of the airport
    1.2 Summary traffic and financial performance
    1.3 Investing in the country
2. *Business overview*
    2.1 Location of the airport
    2.2 History of the airport
    2.3 Site layout
    2.4 Divisional financial summary
    2.5 Organisational structure
    2.6 Other airport service providers
    2.7 Company details
3. *Traffic analysis and outlook*
    3.1 Historical passenger traffic trends
    3.2 Aircraft movements and fleet mix
    3.3 Airline market share
    3.4 Based carrier's operations
    3.5 Connecting vs. O/D passenger mix
    3.6 Destinations analysis
    3.7 Seasonality
    3.8 Peak hours
    3.9 Cargo traffic
    3.10 Management's traffic forecasts
    3.11 Airport contingency planning
4. *Overview of airport site and access*
    4.1 Airport location and development
    4.2 Road access
    4.3 Rail access
    4.4 Assessment of modal access
5. *Aeronautical activities and infrastructure*
    5.1 Runways, taxiways, aprons
    5.2 Passenger terminals
    5.3 Cargo
    5.4 General aviation and VIP facilities
    5.5 Aircraft hangars and maintenance
    5.6 Utilities
    5.7 Navaids
    5.8 Monitoring and controlling systems
    5.9 Telecommunications infrastructure
    5.10 Air transport specific information systems
    5.11 Winter services at the airport
    5.12 Maintenance

Airport PPP preparation 173

6 *Service providers (ramp handling, passenger services, in-flight catering, fuel)*
   6.1 Financial performance
   6.2 Organisational structure
   6.3 Customer analysis
   6.4 Market analysis

7 *Commercial activities*
   7.1 Introduction
   7.2 Duty free and duty paid
   7.3 Food and beverage and other retail
   7.4 Real estate
   7.5 Off-airport real estate competition
   7.6 Ground transportation services
   7.7 Car parking
   7.8 Advertising

8 *Relationship with based carrier/s*
   8.1 Background
   8.2 Status of the agreement
   8.3 Acquisition of based carrier/s' aircraft fuelling business
   8.4 Settlement of based carrier/s' outstanding liabilities
   8.5 Ground handling activity fees
   8.6 Compensation to based carrier/s for on-airport real estate investments
   8.7 Real estate rental
   8.8 Aeronautical charges amendments
   8.9 Net financial impact

9 *Operating and regulatory regime*
   9.1 Country's legal system
   9.2 Legal regulations on foreign investments
   9.3 Country's aviation legislation and policy
   9.4 Regional/supranational regulations
   9.5 Relevant airport legislation
   9.6 National security interests
   9.7 Safety management
   9.8 Security and policing
   9.9 Slot allocation
   9.10 Night-time curfews
   9.11 Noise
   9.12 Protection of national heritage and environment
   9.13 Runway operational restrictions
   9.14 Country's air navigation service provider
   9.15 Building control and permits
   9.16 Customs, immigration, and police
   9.17 Crash, fire and rescue services
   9.18 Fire detection, alarm and extinguishing systems in buildings

174 *Airport PPP preparation*

10 *Management and employees*
    10.1 Organisational structure
    10.2 Board of directors, supervisory board and senior management
    10.3 Terms of employment
    10.4 Employment terms applicable to the management
    10.5 Employee arrangements
    10.6 Pensions
    10.7 Employee data
    10.8 Trade unions and labour relations
    10.9 Training
11 *Legal issues*
    11.1 Taxation
    11.2 Duties
12 *Regional airports*
    12.1 Location of regional airports in the country
    12.2 Summary of regional airports
13 *Financial information*
    13.1 Preparation of pro-forma financial statements
    13.2 Accounting policies
    13.3 Pro-forma profit and loss statements
    13.4 Historical balance sheet
    13.5 Cash flow statement[12]
    13.6 Reconciliation to published annual accounts
14 *Appendix*
    14.1 Glossary
15 *Attachment: Audited financial statements*

---

12 An Information Memorandum should never incorporate a projection of financial results (revenues or costs). A projection of financials might be considered as a grantor guarantee of the accuracy of those financials and at the very least would diminish the negotiation position of the government.

9 Procuring airport PPP

Too great haste leads us to error.
Molière (1622–1673), French theatre writer, director and actor, one of the masters of comic satire, from *Sganarelle*, I. 12

This chapter describes project preparation and implementation. Competitive procurement of PPP involves careful preparation, reviewing risks and their allocation, identifying market requirements and creating a competitive process for selection of the right private partner. In its most basic form, the tender (or bid) process involves a party offering a project to the market and asking for bids from parties interested in performing the project, or some part of the project. Tendering (or bid) procedures are meant to achieve efficiency, manage costs, maintain quality, encourage expediency and maximise value-for-money. PPP transactions take time to prepare, and need the attention of experts to ensure that risks and financing are managed properly and efficiently and that they are taken to market in a form and manner designed to attract as many high quality bidders as possible and thereby keep costs down and improve delivery.

The preparation stage can take time. Decision makers should expect at least six months before bidding stage. Starting the bidding too early, or selecting a specific operator without any competition, ends up taking longer and costing more on average. Experience has shown that time and money spent on open, transparent, competitive bidding increases the chance of success of the transaction. Once bidding starts, decision makers should anticipate at least six months to contract signature and an additional six to twelve months to financial close (if needed).

Management contracts are less complex than concessions and require less preparation, time and effort. However, a strategy for air transportation and master plan may still be needed, especially if the management contract is seen as a first step towards increased private participation.

Once a project is selected, established through a masterplan and verified through a feasibility study, it is time to prepare the transaction and take the project to market. This process is expensive and time consuming but can be used to attract a larger number of high-quality bidders, increase access to financing, mobilise more investment and improve the likelihood of success of the project.

The procurement rules that apply to the project will follow the national and/or state/local legal/regulatory regime. There may be special rules for PPP projects, as traditional public procurement differs significantly from PPP procurement due to duration, the nature of the relationship with the investor and the type of financing used. One might consider PPP procurement as a hybrid between public procurement of capital investment and the privatisation of a public asset. The applicable procurement regime may also be influenced by policies and practices in the sector or imposed by financiers, for example bilateral and multilateral financiers will require a particular approach to procurement.[1] In some cases, advice will be needed to coordinate compliance with all of the different applicable procurement regimes.

Figure 9.1 maps out the steps in the procurement phase.

9.1 Competitive processes

Governments often receive proposals directly from private investors. These proposals can be a good source of innovative ideas for the government and can help governments identify new project concepts. However, unsolicited proposals

---

1 See for example the World Bank's procurement rules at www.worldbank.org/procurement, which require open, competitive procurement in an effort to achieve economy and efficiency. Any investment financed by the World Bank must comply with these procurement rules.

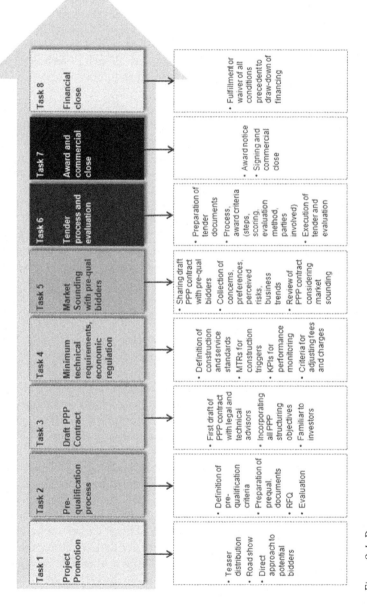

Figure 9.1 Procurement

are difficult to manage and can be a source of significant mischief. Serious technical capacity and experience with PPP is needed to manage them well.

Direct negotiations generally take longer, are more expensive and are more likely to fail than projects procured through competitive processes.[2] Directly negotiated arrangements are also more vulnerable to challenges by new governments or opposition groups – without the validation of a transparent, competitive process, direct negotiations are more susceptible to claims of bias, corruption, incompetence and inappropriate use of government resources.

Where permitted, the circumstances allowing the award of the project without competitive procurement should be limited, for example by applying the following criteria:

- Where the project is of short duration and a small value, such that the added efficiency of a competitive process is outweighed by the cost of the process
- The project is critical to national defence or national security, and the competitive process would require disclosure of sensitive security information that cannot be managed safely
- There is only one possible source of the services (due to the skill set of the provider or exclusive intellectual property rights)
- Where there have been repeated efforts to implement a competitive process, but with no success, yet there is one party willing to undertake the project on the same terms that failed to attract competition (see Box 9.1).

Box 9.1 If at first you don't succeed

After failed attempts to procure an investor for the first greenfield totally developed by the private sector in Ecuador right after the government's financial default, a respected group of investors supported by the Canadian government and backed by the technical expertise of Houston Airport approached the city government of Quito. A structured negotiation process ensued between Quito and the group of investors, with a Swiss Challenge mechanism to allow other interested parties to engage. The resultant concession was a clear success.

The city of Chicago launched a procurement process in 2009 for the Chicago Midway airport. The bidder failed to secure financing, and therefore the process was cancelled. In 2013, the city proposed a shorter lease term and revenue sharing arrangement. The transaction terms resulted in one of two bidders withdrawing. The city reappraised its position and cancelled the bidding process, as it was unlikely to obtain a reasonable arrangement with a single bidder. This was a sensible decision (and in line with IATA's 2018 guidelines on airport ownership and regulation). The

---

2 See inter alia: UK National Audit Office, "Getting value for money from procurement" (2001); Henrik M. Inadomi, *Independent Power Projects in Developing Countries* (2010).

> government should not be forced into an unviable situation; if the bidding process needs to be cancelled and redesigned, then government needs the right and mandate to do so.

Whenever it is proposed to award a project without competitive procurement, the mechanism to apply for such a waiver should be managed by an appropriately high-level authority. The decision process should be made public and transparent to allow other stakeholders to comment if they have issues, and there should be a mechanism for those disgruntled stakeholders to appeal against the decision. These mechanisms help protect the decision and the project from vulnerability to legal and political challenge.

Where competition is not possible or practicable, legislation often provides for market testing (see Box 9.2) to ensure that the pricing and terms agreed for a directly negotiated project to meet market standards (consistent with what the government would have achieved through competition). A robust, independent feasibility study is invaluable in such circumstances.

> **Box 9.2 Benchmarking**
>
> Where a project is not subject to competitive pressures, or that competition is insufficiently robust, the government should submit that project to benchmarking to verify that the price represents best value as compared to similar airport projects and in similar countries. This can be a challenging process where equivalent projects are not readily available, or where relevant information is not available.

Some countries reject unsolicited proposals outright, providing no benefit or compensation to those offering such proposals, particularly countries without the resources and sophistication to manage unsolicited proposals. This offers a robust method to avoid the complications and dangers of unsolicited proposals but also deprives the government of any advantages.

Unsolicited proposals are often attractive to contracting authorities as the proponent offers to prepare a feasibility study for the project at no or low cost, in exchange for the award of the project. However, the contracting authority cannot rely on the feasibility study. Even if the study is provided in the detail and with the rigour that the contracting authority would apply to its own study, the proponent is naturally biased to show the project is viable and should use the technology and methodology familiar to the proponent. The study will need to be reviewed and verified by independent advisers before the contracting authority can rely on it.

The provincial government of Bali, Indonesia, has received an unsolicited proposal for a new airport in the north of the island, as a complement to Ngurah Rai International. The unsolicited proposal included technical and economic studies that were significantly biased, overstating the traffic prospects, underestimating the investment costs and misrepresenting the economic benefits of the project to the economy of the province.

Mechanisms have been developed to encourage unsolicited proposals, while also ensuring that competitive tendering is used, where possible, when selecting the best investor.[3] These mechanisms involve a careful review of such unsolicited proposals to ensure they are complete, viable, strategic and desirable. The project is then put out to competitive tender, with the proponent of the unsolicited proposal receiving some benefit, for example:

- Proponent pre-qualified automatically
- A bonus on the proponent's scoring in the formal bid evaluation (i.e. additional points allocated to the proponent's total score when its bid proposal is evaluated)
- A first right of refusal, enabling the proponent to match the best bid received (also known as the "Swiss Challenge"), in some cases only where the proponent's bid score is within a defined margin of the best bid
- The right to automatically participate in the final round of bidding, where there are multiple rounds of bidding (the "best and final offer" system); and
- Compensation paid to the proponent by the government, the winning bidder or both.

Box 9.3 Unsolicited proposals in Colombia

The Ministry of Transport, DNP (the planning agency) and the MoF enacted detailed regulations regarding the acceptance of unsolicited proposals from the private sector.* If accepted as a viable project, an unsolicited proposal must then go through a competitive, open procurement.** The proponent participates in this selection process like any other bidder. If the proponent's bid is not selected, however, then the winning bidder must reimburse the proponent for certain of its expenses, as approved by the responsible government agency prior to the start of the tender process. In such a case, the proponent is responsible to the winning bidder for the quality of the relevant studies.

* MOT Decree No. 4533 of 2008.
** See requirements of Laws 80 of 1993 and 1150 of 2007.

---

3 See here Hodges and Dellacha, *Unsolicited Infrastructure Proposals: How some countries introduce competition and transparency* (2007).

The unsolicited proponent is often viewed as having an unfair advantage, so any preference given (such as a right of first refusal or bonus during bid evaluation) may stifle competition.[4] A more robust approach is to use competitive tendering, but without any advantage to the unsolicited proponent. Instead a fee is paid to the proponent if he does not win the bid, as compensation for the value added by his efforts to develop the project (see the example of Colombia in Box 9.3). The fee should be sized to reflect actual benefit of the proposal to the government.

9.2 Pre-qualification

The bidding process is generally lengthy and costly, for the bidders and for the contracting authority. In order to manage the cost and time outlay, the contracting authority may wish to prequalify those parties most likely to provide an attractive bid. The prequalification process:

- Avoids the time and cost of managing bidders who do not have the fundamental qualifications or financial substance that would enable them to undertake the project
- Encourages good bidders, who would prefer a smaller field of equally qualified competitors. Low-quality bidders are more likely to low-ball a bid and their presence amongst short-listed bidders may scare off high-quality bidders
- Enables dialogue with a specific number of potential bidders, improving market sounding, verifying PPP agreement terms and conditions with the prequalified bidders, and exploring concerns and risks. Without prequalification, there is no discrete subset of bidders for such interactions
- Allows governments (politicians) to demonstrate the advantages of a robust process and the quality of bidders attracted to the investment. This often allows government more time to implement the procurement process well.

An additional benefit is that it helps governments (politicians) to calm anxiety in terms of being able to show in advance the process and the prospects of good bidders, allowing the process to take the time it requires.

The government must decide what is expected from the bidder, or the bidding consortium, during the prequalification stage. Ultimately, this is a transportation policy decision.

Some governments expect the bidder to be an airport operator with proven experience in management. However, it might not be necessary that the airport operator be an equity holder, but rather to be engaged with the investing bidders through a technical service agreement (TSA). The equity participation of an airport operator does not necessarily mean real hands-on managerial responsibility

---

4 This is not the case in countries like Chile, with a sophisticated regime that gives confidence to other bidders that the proponent will not have an unfair advantage in the process.

in the day-to-day business, such as appointing the top managers. An ideal TSA would ensure that:

- Executive managerial positions are handled by personnel that have spent a number of years in a managerial position, for example, a manager of operations at the investor's company
- A team of managers will train local staff on a regular basis
- Some of the managers of the new PPP company will be brought to the investor's home airport to train, for significant periods of time
- Processes, systems and procedures should be transferred as part of knowledge/ know-how/technology.

One additional challenge with respect to the investors is the way in which they may qualify as "operators". For example, one airport operator may be using a subsidiary unit to bid for the PPP, which is not exactly the company that manages the airport at their home. The prequalification must contemplate this difference and accept the subsidiary as good as its parent company. Another example could be that the operator partner has a partial share at their home airport, but is in fact the actual managing partner. Then the question is how to demonstrate that this particular partner is in fact the one that is responsible for the day-to-day operation and not a passive investor, for example at least two of the CEO, CFO, COO, chief marketing officer and chief commercial officer are appointed by the investor.

For some PPPs, the importance of having a local investor is crucial for the success of the project. Some governments may even introduce the nationality issue as a requirement, and may even preclude some nationalities on the grounds of national security. The political connections and local knowledge could be essential in some countries, and less relevant in others. For example, attempting to bid in Mexico without a local partner could be inefficient. However, the French-Italian consortium that won the bid for Santiago de Chile did not involve a local partner in their consortium. In some countries, it could be hard to find local partners complying with some of the prequalification criteria.

It could also be that the grantor may wish to request a construction company as part of the consortium. However, it should be noted that the nature of the project may require that the construction company only be involved during the initial phase, while at many airports today the refurbishment and expansion is an ongoing and continuous process that could last throughout the entire life of the PPP.

Box 9.4 Prequalification criteria for an airport operator

Each sector and project has its own specificities. For example, prequalification criteria may include:

- Specific experience in operating an airport with a certain volume of traffic, for no less than a defined number of years

- Level of net worth in excess of a set amount
- Recent experience managing the construction and operation of a project of similar size and complexity in a similar market
- Recent experience raising similar amounts of debt and equity.

The criteria may also consider other restrictions, for example exclusion of:

- Air carriers, or of companies owned by air carriers
- Operators of airports located close to the site, e.g. within a specific radius (which could create a possible conflict of interest)
- Foreign state-owned entities (SOEs)/ airports (for national security issues, in particular where the airport has any military use).

The prequalification (PQ) documents should include:

- Brief project description (typically through an Information Memorandum or a *teaser*), project size and beneficiaries, broad implementation frameworks and timeframes, estimated cost, etc.; and
- Evaluation criteria, including the minimum criteria for prequalification.

9.3 Bidding

The contracting authority provides the prequalified bidders with tender documents (also known as bidding documents or a request for proposals (RfP)) and access to relevant data. Bids received will be evaluated against specified criteria. The criteria need to be described thoroughly in the bidding documents to help bidders understand the contracting authority's needs, and to improve the quality of bids received.

Even in the most sophisticated markets, creating investor appetite can be a challenge. The procurement process should put the contracting authority into a strong negotiating position if there is only one bidder, limiting the opportunity of that bidder to hold the contracting authority hostage. The contracting authority needs to be prepared to start the bidding process over if it is not happy with the bidder's proposal. Where a single bid scenario is encountered, benchmarking of the bid may be a useful mechanism to help the government understand if it is getting good value, and to help reassure other stakeholders that the lack of competition is not a fatal flaw in the process.

Pre-qualified firms will receive bidding documents, which set out a detailed project description and the basis for the bid, including bid process rules and draft contracts. Extensive dialogue with prequalified bidders is highly recommended to help them understand the government's requirements and for the government to understand bidders' concerns. In order to prevent perception of bias or unfair advantage, communication needs to be managed carefully – structured and transparent. Following dialogue with bidders, bid documents will then be revised and bids will be submitted. Sufficient time should be given to bidders to consult the documents provided (including traffic data, site details, use of land, existing contracts, regulation in place, etc.), survey the site, and prepare a more detailed proposal.

Bids may have technical, financial, and legal elements with the following requirements:

- A technical solution compliant with the airport conceptual masterplan and specifications provided. This could be a fixed technical solution imposed by the government, or a solution based on output-based specifications. The technical proposal should also include the proposed management team that will be taking over the airport, including the qualifications of each team member and their responsibilities
- An airport financial proposal indicating how much government funding, investment, or guarantee is needed or how much of the project revenues will be shared
- An airport financing plan showing how and from whom the bidder intends to mobilise debt, equity and other financial instruments to fund the project and how much due diligence has been completed. This will include a financial model showing the bidder's assumptions.

The technical proposal could be either scored or pass/fail. While the former could lead to subjectivity in evaluation by the reviewers, it allows bidders to provide more innovative technical proposals, rather than just good enough to pass. The technical proposal must be binding, such that the bidder commits to carry out what they have proposed.

A non-binding technical proposal can be an interesting intellectual exercise, but does not commit the bidder legally and is therefore not useful in the procurement process.

It is also common to request, as part of the financial proposal, a financial model that supports and explains the proposed financial offer. The financial model is important to avoid unrealistic offers or errors. The financial model needs to be accessible to the contracting authority, so that formulae and calculations can be reviewed and tested.

A negotiation phase may follow the selection of the preferred bidder (in particular when the bid is not underwritten – i.e. when the lenders are not fully committed to the project yet, and therefore need the opportunity to influence the concession agreement) and with the objective to reach financial close (when financing is secured and available). Even when the concession agreement is signed, the government may need to anticipate additional time to secure financing. This negotiation should be transparent and structured in a way to reject any material change to the bid documents.

If the government cannot conclude negotiations with the preferred bidder, it should allow itself to proceed to the next best bidder. However, this can be time consuming and expensive. Lenders are expected to finalise their due diligence once the preferred bidder is selected, but bidders whose lenders have completed more of their due diligence and are closer to underwriting the project (beyond the minimum levels required) should be rewarded accordingly in the bid evaluation process. It may be that the government requires bidders to bring fully committed lenders, or at least any lender comments, into their bid, and therefore limits or excludes any further negotiation post-bid. Some countries do not permit material changes to the project post-bid.

The bidding process should include the following:

- Select the right tender board members, with a mixture of technical, financial and legal expertise, sector knowledge and PPP experience
- Establish rules and process for tender board to evaluate proposals and apply criteria, including format for receipt of proposals, proposal evaluation scoring, and writing up board decisions. The report shall summarise the selection and recommendation of the bidders selected and the reasons for proposals rejected
- Assessment of the entire proposal, and extent of compliance with technical aspects; these could include some of the following:
  - Technical standards, including design criteria, timing for completion, aesthetic qualities and service standards
  - Performance standards
  - Proposed management team and know-how / technological transfer
  - Environmental and social requirements
  - RFQ qualification requirements
  - Bidder's monitoring methodologies and reporting proposal
  - Workability and security systems
  - Disaster/emergency management systems (in particular to comply with regulatory requirements)
  - Bidder's procurement plan and procedure for managing sub-contractors
  - Financing, including how much progress has been made by the financiers towards financial close.

Dialogue between the government and the bidders needs to be structured to avoid wasting time and duplication of effort, and to maintain the government's position in the bidding and negotiation process. The dialogue between the government and bidders focuses on a few key issues typical of airport projects.

Bid criteria for an airport PPP usually include the following:

- *All documents submitted, signed, notarised and official* as required by the bid documents, for example affirmations of corporate standing, attestations of good standing, bid bonds, and so forth
- *A technical solution compliant with the airport conceptual masterplan and with specifications provided.* This could respond to a fixed technical solution imposed by government, or propose a new solution based on output-based

specifications. Government-defined technical solutions help improve certainty, but constrain the bidder's ability to innovate and be creative in its bid solution. The solution may also have an artistic component of aesthetic value. The technical solution may be graded higher if it provides for earlier completion, or for longer life-cycle assets. However, adding these variant factors as part of the selection criteria may reduce objectivity and impartiality during the proposal evaluation

- A *legal solution compliant with bid documents*. The government may allow bidders to proposed amendments to the PPP agreement. If amendments are proposed, bidders will lose points if the amendments place more risk on or increase the cost for government. Evaluation of the financial solution can become a very subjective exercise and the government should provide a clear regime for assessing the implications of proposed changes and the number of points to be deducted. The government must establish whether it is required to accept amendments proposed in the bid, or if these amendments are simply proposals the subject of negotiation after bid award (if this is legally permissible). Generally speaking, the bidder should not be given the right to propose amendments to the bidding documents; this is too hard to evaluate, and generally causes additional complication during commercial and financial close. It is much more advantageous to identify such changes during market sounding and implement them in the draft PPP agreement for all bidders
- A *financial proposal* indicating the extent of government funding, investment, or guarantee needed or the share of project revenues. For cost-sharing arrangements, the one asking for the lowest subsidy will earn more points. Where revenues are shared, the bid proposing the highest revenue share will earn more points
- A *financing plan* showing how and from whom the bidder intends to mobilise debt, equity and other financial instruments to fund the project and how much due diligence has been completed – all supported by letters of intent or commitment from those different financial investors and lenders. The financing plan will include a financial model showing the bidder's assumptions. The plan informs the government of the time needed to reach financial close and the likelihood of failure to mobilise funding. A less convincing financing plan, e.g. with little lender due diligence completed, will be graded lower. Alternatively, the government may require an underwritten bid (where lenders commit unreservedly to provide financing), although this will require careful market analysis as underwritten bids are difficult and expensive for bidders. Letter of intent or commitment from possible financiers could be required as an alternative to an underwritten bid.

Tender documentation would indicate the relative weight of these elements, for example 60% technical and 40% financial. This weighting reflects the fact that the least expensive bid is not always the best; airport projects involve a careful balance of technical, operational and commercial context, requiring a bidding consortium to offer solutions to all key challenges and able to respond to change and crises. Governments should measure the value-for-money represented by each bid, the technical and financial standing of the bidder and the likelihood that the project will be a success if that bidder is selected at the price bid.

Most airport PPPs involve an important transition period, when functions are gradually transferred to the private operator. Transition avoids any disruption in services and allows a softer adjustment for staff and service providers. During transition, the government will need to:

- Develop a check list of outstanding issues
- Process asset and staff transfers
- Verify data provided in the bidding documents or at the time of bid to confirm accuracy and completeness
- Finalise adjustments (including of contract).

Creating investor appetite can be a challenge; even before the 2008 financial crisis, some 30% of PPP projects in the UK only received two bids.[5] The procurement process should put the contracting authority into a strong negotiating position if there is only one bidder. The contracting authority needs to be prepared to start the bidding process over if it is not happy with the bidder's proposal. Where a single bid scenario is encountered, benchmarking of the bid may be a useful mechanism to help the government understand if it is getting good value, and to help reassure other stakeholders that the lack of competition is not a fatal flaw in the process.

9.4 Award/negotiation/financial close

Once bids are received, the contracting authority will evaluate compliant bids and select the preferred bidder. The contracting authority will negotiate with the preferred bidder any open issues (to the extent permitted by the bid documents or by law), finalise the commercial and financial arrangements, award the project, sign the concession agreement and other key contracts (subject to the conditions precedent discussed below), and reach financial close. More than one preferred bidder may be selected for additional rounds of competition, for example through best and final offer (BAFO – see Box 9.5) or competitive dialogue. Additional rounds need to be carefully managed, to maintain transparency, avoid any perception of favouritism or corruption, and limit the added cost and delay such a process implies.

---

5 From the National Audit Office (UK), "Improving the PFI Tendering Process" (March 2007).

> **Box 9.5 BAFO**
>
> The contracting authority may choose to include additional stages of competition, for example reducing the competition to two bidders who will then be asked to further refine their bids and submit a best and final offer (BAFO), further to which the contracting authority chooses the preferred bidder. This process allows the contracting authority to use the available competitive pressure to further motivate bidders, and possibly obtain firm financing commitments.

Lenders will not be finally committed to the project until financial close is achieved. Before financial close, lenders will want to confirm that the risk allocation for the project is "bankable", a general term referring to the level of comfort that a lender will require from a project given the context of the project.[6] The lenders will then agree with the project company and the government a list of conditions precedent (CPs) that must be satisfied before the lending arrangements become final, and before first drawdown can be made.

The signing of a PPP contract does not necessarily lead immediately to the beginning of construction. The project company will likely receive some (20–50%) of its total investment needs from its private owners as equity, and it will want to raise the majority of its finance (50–80%) as debt from lenders. However, the lenders will insist on performing their own detailed due diligence on the project before providing this level of debt. This process can take three to six months to complete (and up to a year for larger, complex projects). The financiers may start due diligence before project award. If they complete due diligence before bidding and the offer of finance is firm, then the bid is "underwritten". Further assessment by the lenders may be required after award but before the offer of financing becomes firm. This involves a second event called "financial close", at which point all conditions precedent have been satisfied or waived and first drawdown of debt can be made.

Reaching financial close can be a very demanding process. Financial close is part of the relationship between private investors and their lenders, however governments or contracting authorities may need to be involved, for example where the lenders

- Require additional information from the public sector
- Wish to propose changes to the project agreements
- Require direct agreements with or other support from the government/contracting authorities.

---

6 See Delmon, *Private Sector Investment in Infrastructure* (2009).

In order to establish its role in ensuring timely financial close, the contracting authority may want to:

- Give the project company a maximum period to achieve financial close, generally twelve months from contract award
- Ask to be informed about progress of negotiations, with weekly updates and more detailed periodic reporting, e.g. monthly. The contracting authority may also want access to the lenders to assess whether the project company is proceeding diligently towards financial close
- Respond diligently to requests from lenders for additional information
- Stand ready to enable/encourage timely financial close, including where the investors do not have prior experience in raising project financing for PPP projects, the project appears too risky for lenders to finance, changes in the consortium members are required or other challenges arise.

While the contracting authority will not want to take over financing responsibility, failure to reach financial closure is a costly experience for all parties, and will merit efforts by the contracting authority to support the process.

10 PPP agreements

> Don't talk unless you can improve the silence.
> Jorge Luis Borges (1899–1986), Argentine writer, poet, critic, translator and man of letters

PPP structures for airports are ultimately flexible. The models discussed in this book are provided as an example of the different approaches that one might take to PPP structures, but should not be considered exhaustive.[1] This chapter will

---

1 The World Bank's PPP Legal Resource Center provides check lists and sample laws, regulations, concession agreements, contracts and other useful materials for practitioners in airport PPP, amongst other sectors. See www.ppp.worldbank.org.

discuss some of the key issues (10.1) and risks (10.2) addressed in PPP agreements, applicable generally to the different types of PPP, and the contractual provisions used to manage these issues and risks (10.3). Instead of endeavouring to dissect contractual structures and risk allocation for every possible PPP structure, this chapter will discuss key issues that arise in any airport PPP project.

10.1 Key issues

The PPP Agreement will address a number of key issues that form the basis for a good airport PPP, including:

10.1.1 *Legal/regulatory context*

The contracting authority must have the right to implement the project, including the power to delegate project activities to the project company and must be duly authorised to undertake the PPP agreement, often with the approval of some combination of the airport authority or state owned airport company, the Civil Aviation Authority, the Ministry of Transport, the Ministry of Finance, and the local municipality. The provision of air navigation services remains under the scope of the government, either at a dedicated authority or in some type of state-owned corporation. Functions like customs, immigration and security also remain with the government.

The applicable law will set construction and operation standards, for example environmental standards, safety and security standards, and operating requirements. The PPP agreement will establish a regime for changes in law, to allocate this risk as amongst the contracting authority and the project company.

Air transport is heavily regulated, in particular safety, security, market access, slot rights at airports and economics. The project company will not have complete freedom in its operation of the airport, even if the PPP agreement says it does. The PPP agreement should be developed in close collaboration with the regulator (financial and technical) and consistent with the regulatory framework. The concessionaire will need the right to challenge the regulator's decisions, and may need protection from the contracting authority where the regulator makes a decision contrary to the PPP agreement (e.g. where the PPP agreement establishes technical standards to be implemented, but the regulator requires a different set of standards that will cost the project company more money to implement, or where the regulator sets landing fees at a level lower than agreed in the PPP agreement; the PPP agreement needs to establish what if any compensation the project company will receive from the contracting authority).

10.1.2 *Scope*

As discussed in Chapter 6, the PPP agreement will set the functional scope of the concession and its term. The project scope will need to address a number of practical activities in and around the airport that may require investment and may provide additional revenues for one or the other of the project parties, for example:

- The provision of services that will be transferred to the project company, for example the responsibility to:
    - Collect fees and charges
    - Invest and maintain the related infrastructure
    - Provide necessary infrastructure for the operation and movement of aircraft
    - Deliver the landside facilities comprising the terminal buildings, facilitating the processing of passengers from air to surface transportation, and vice-versa
    - Provide facilities to other government entities necessary to process passengers (passport control, customs, police, health, etc.)
    - Deliver a specified level of service (normally linked by the right to levy fees and charges on users, typically collected directly)
- Airport access and links to the city. The airport might need a specific link to a nearby road or highway, particularly when building a new greenfield airport or a new terminal. The airport may reserve the right to grant access to specific providers of ground transportation, depending on applicable local legislation
- The operation of vehicle parking in and around the airport is typically included within the scope of the PPP, often granted on an exclusive basis. The operator may be able to prevent outside operators from picking-up and dropping passengers from off-airport parking facilities
- The provision of services to airlines may be the prerogative of the project company, or may be allowed from other providers on a competitive basis, for example the supply of fuel to aircraft, in some cases reserved for one or a specific number of vendors. The European Union has a specific regulation with respect to competition for specific services, depending on the volume of traffic of the airport. It is not unusual to find PPP where the provision of a particular service is not included within the scope.

The scope set out in the PPP agreement will be supported by specific minimum technical requirements that permit the contracting authority to test whether the project company complies with its obligations. A penalty mechanism will help the parties focus on the most important performance criteria and establish a regime to identify defaults early, in order to resolve them and avoid disputes.

The PPP agreement should establish mandatory investments to ensure that the project company complies with agreed safety and security standards, while delivering a minimum level of service. This will be linked with the time for completion of different phases of the work, and design specifications, in particular.

The PPP agreement will establish responsibility for the design specification (often provided by the contracting authority), for example the capacity of the runway system to operate a specific number and size of planes, the capacity to accommodate aircraft based on the number and type of parking stands, the amount of passenger throughput during a given period, under specific level of service conditions, etc. Investments should allow the operation under specific standards and regulations, including civil aviation, environmental (noise and pollution standards) and fiscal (e.g. custom bonded areas). The PPP agreement will need to establish liability for errors in the design specifications and a regime for interfaces between the planners (engineers, designers, architects) and the project company.

The project scope established at the time of the award of the PPP may need to adjust to events and circumstances that arise but were not contemplated in the PPP agreement. The PPP agreement will therefore address scope expansions, how these are to be defined and priced, whether they are to be the subject of a separate tender process, will be subject to a separate agreement or part of the same arrangement and the extent to which they can be pre-defined (often known as provisional sums). These provisional sums relate to short term capital expenditure. Beyond a five-year span, the referential master plan sets out minimum technical requirements (MTRs) and levels of service (LoS) that apply to the operator, but will not define works and sums. Every five years the operator should present a detailed master plan in compliance with these MTRs and LoS. Once approved, the master plan sets the specific works to be carried out over the next five years.

In order to protect the demand profile of the project, the contracting authority may need to provide the project company with exclusivity over the scope, e.g. where the government undertakes to ensure that other airports are not built within a certain geographic region that reflects the relevant catchment area or the nature of the airport market. Exclusivity may not apply to certain activities, such as airports for private / non-commercial services, military bases and airports with restricted operations to particular routes (e.g. domestic and regional operations vs. international). In some cases, exclusivity may include other competing technologies, such as high-speed rail links.

10.1.3 Site access

The project company will need access rights to the site, to design, build and operate the facilities. The timing of site access will be critical to delivery by the project company. The PPP agreement will set out in detail the site access rights to be granted by the contracting authority to the project company, what site access the project company must provide, the timing of such access, and any damages payable by the contracting authority if access is delayed. The PPP agreement will also define what environmental or other liabilities are to be borne by the project company in relation to the site, and what insurances are to be procured to address these risks.

> Note: Site access can create serious liabilities for government where timing agreed in the PPP agreement is overly optimistic. The cost of delay to the project company can be significant, therefore penalties imposed on the government for delay of land acquisition will be commensurate. There may be limits within the law or politically as to the type of property rights that can be granted by the contracting authority to the project company and whether the project company will have the right to sub-let/ franchise parts of the site to sub-concessionaires.

At the end of the project period, the site will be handed back to the contracting authority. The PPP agreement will establish handover requirements. See Section 11.6 for further discussion of handovers.

10.1.4 Personnel

PPP airport operations often involve a large number of personnel. Where existing airport operations are to be transferred to the private sector, airport personnel may also need to be transferred from the public sector to the project company. An extensive consultation process is needed, led by the government and the project company. The government may require that a proportion of airport staff be recruited locally.

Where employees will transfer to the project company, the parties (possibly including employee representatives or trade unions) will need to agree the benefits package associated with those new positions, for example pay arrangements, bonuses/ performance incentives and pension entitlement. It may be that public employees are seconded to the project company, in order to maintain their public sector benefits entitlement. The government may want to limit the number of staff that the project company is allowed to lay off or make redundant in the initial period of the project. This will usually be agreed with the labour unions and the local community.

The PPP will need to establish what happens to personnel on hand-back of the concession, including seconded public personnel and those hired by the project company, to protect the personnel and ensure continuity of services.

> Note: Consultation with employees needs to be prioritised and implemented early in the project design process. Failure to consult can delay project implementation, increase costs or even result in complete project failure.

10.1.5 Existing business

An airport project that involves existing operations, with existing business, will need to consider a number of key issues:

- Existing contracts will need to be *novated* to the project company, for example agreements with airlines, fuel supply agreements, ramp handling companies, in-flight catering companies, etc. An assessment is needed of these agreements, the extent to which they can be transferred, the amount of control given to the project company, as compared to the government, and any associated costs. It is not unusual to see state-owned operators rushing to close deals with service providers or retailers just before the PPP is implemented. Since this practice tends to be motivated by opportunism it is always unfortunate as it diminishes the value of the operation, since most deals are not necessarily beneficial for the grantor. Whenever possible, sealing new agreements prior to the implementation of a PPP should be prevented
- Existing personnel, as discussed above, will need to be transferred to the project company or retrenched
- Existing business relationships (whether or not established by a long-term contract) will form the basis for the existing business but will need to be reviewed by the project company. Some of those business relationships may not be comfortable engaging with a private entity, for example state-owned enterprises. In other cases, the project company will have its own business relationships and will want to replace those of the government. This may create additional tension between the project company and the contracting authority to agree how to modify existing arrangements to meet project company requirements while respecting pre-existing government arrangements. An airport concession in Argentina left the monopoly of ramp handling services with a company owned by the Ministry of Defence, causing continuous tensions with the landlord and users[2]
- Investors will be concerned that the existing operations may have caused some environmental damage, or created further environmental risks, that would be borne by anyone taking over the existing business. The PPP agreement will need to allocate environmental liabilities, existing and future. The project company may only be responsible for environmental liability arising after the date of the contract, latent defects (those caused before transfer but whose impact is not perceived until after contract) or some or all of the remediation of existing damages. It should be noted that it may not be possible under law to assign some of these liabilities.

10.1.6 Revenues

Project revenues can be derived from a number of sources, including airport fees and charges, access charges from sub-concessions, parking, secondary development, etc. The PPP agreement will need to set out the nature, terms and allocation of those revenues. The determination of fees and charges can be particularly challenging as they are often regulated. Even if unregulated, fees and charges may be politically sensitive. See Chapter 5 for further discussion of airport revenues.

---

2 As of early 2019 the Ministry of Transport has finally decided to liberalise the ramp handling services (*Boletín Oficial Argentina*, Decreto 49/2019 – DECTO-2019-49-APN-PTE – Servicio de atención en tierra de aeronaves, 15 January 2019).

The project company may need to pay a concession fee to the contracting authority, usually some combination of a one-off fee and periodic payments (often a share of revenue). The timing and conditions for payment, including penalties for late payment, will be set out in the PPP agreement.

10.1.7 Finance

Sourcing finance is generally a risk/task allocated to the project company. In order to fulfil this task, the project company may need to create security for the benefit of lenders over project assets, revenues, insurance, etc.[3]

In some cases, the government will share certain financing risk, for example, the bidders will submit bids based on the cost of financing at the time of bid. The government may share in the risk of market shifts, e.g. base rates between bid date and financial close. Governments may also share in refinancing risk, for example where local currency risk is only available for short tenors. Governments may also want to share in refinancing gains, where refinancing results in a significant improvement in project profitability.

One of the key challenges of financing PPP projects is foreign exchange risk – where the revenues of the project are in a different currency than the currency of debt. Airport projects are in an advantageous position, mobilising foreign currency revenues, e.g. from international aeronautical revenues and duty-free retail. Where foreign exchange risk is a challenge for an airport project, the government may decide to share the foreign exchange risk.

10.1.8 Dispute resolution

PPP projects have characteristics that lend themselves to recurrent and often debilitating disputes involving parties from a variety of legal, social and cultural backgrounds;[4] they represent long-term, complex commercial and financial arrangements, which may require renegotiation to resolve. Failure to address such disputes early can have a devastating impact on a PPP project. Conflict management, dispute resolution and mechanisms to manage renegotiation are essential elements of the PPP agreement. More important is the attitude of coordination and collaboration. Where an issue or conflict arises the parties must be committed to communication and collaborative resolution as early as possible. Please see Section 11.5 for further discussion of dispute resolution.

---

3 See Delmon (2009) for more.
4 Straub, Laffont, and Guasch, *Infrastructure Concessions in Latin America* (2005). From the PPP database (preliminary figures): for 2003, 34% of contracts (by investment amounts) in the water sector were classified as distressed and 12% were cancelled; in transport 15% were distressed and 9% cancelled; while in energy 12% were distressed and 3% cancelled. See http://ppp.worldbank.org.

10.2 Risk allocation

A successful project must benefit from workable, commercially viable and cost-effective risk sharing. Given the differing interests and objectives of the parties involved, effective risk allocation will be an essential part of the drafting of the project documents and an integral part of the project's success.

Risk management based on efficiency[5] is, of course, an ideal, a goal. In practice, risk tends to be allocated on the basis of commercial and negotiating strength. The stronger party will allocate risk that it does not want to bear to the weaker party. This scenario does not necessarily provide the most effective and efficient risk management.[6]

In most conventionally financed projects, it is accepted that certain risks (such as market risk, certain political risks and completion risk) will be allocated to the project company. For bearing such risks, the project company is compensated by higher returns on its investment. The project company then allocates project risks to its sub-contractors (e.g. construction risk should be allocated to the construction contractor). The effort to transfer all project risk to these sub-contractors is known as "back-to-back" risk allocation. Complete back-to-back risk allocation will result in the transfer of all project risk assumed by the project company to the other project participants. Rarely, if ever, will a PPP project achieve complete, absolute back-to-back allocation, although the most developed PPP markets like the UK achieve something close to it.

The following risks are of concern to parties, in particular in relation to the potential for increase in costs, reduction in revenues or delay in payment.

10.2.1 Political risk

The contracting authority may accept a certain amount of political risk, as the sole party who may be able to influence its advent and mitigate its effects. Political risk includes:

- *Changes in law or regulations*, in particular the risk of discriminatory changes in law (those changes which are specific to the sector involved, private financing of public projects generally or the project itself) and changes in technical parameters through permits, consents or import licenses. With the change of government in Ecuador in 2007, the city government of Quito

---

5 An oft-quoted approach to "efficient" risk allocation places each risk on the party best able to manage that risk. While a useful rule of thumb, this is a gross simplification. (See chapters 1–3 of Delmon's *Project Finance, BOT Projects and Risk* (2005).) For example, risk also needs to be borne by the party that has an interest in managing it proactively; has or will obtain the resources needed to address risk issues as and when they arise (the sooner the better) in a manner intended to reduce their impact on the project; has access to the right technology and resources to manage the risk when it crystallises; can manage the risk at the lowest cost; and delivers value for money.
6 See Business Roundtable US's project report on *Contractual Arrangements* (1982).

forced the concessionaire of the airport into a renegotiation process, changing some of the original parameters of the concession agreement
- *Expropriation*: it is a basic principle of international law that a sovereign government has the right to expropriate property within its territory for public purposes, but must compensate the owner;[7] Venezuela expropriated the airport of Porlamar, General en Jefe Santiago Mariño in Isla Margarita, while it was operated by Zurich Flughafen and IDC Gestión e Ingeniería (Chile). The ICSID international court ruled this expropriation illegal and awarded compensation
- *Decisions from the regulator that differ from the PPP agreement* (e.g. where the regulator sets service level, safety or security requirements at law that are different from the PPP agreement). The contract between the operator of the three major airports in Bolivia (Abertis Infraestructuras, from Spain) and the Bolivian government left open key issues related to the regulation of these airports. Failure to reach an agreement led to the nationalisation of the airports and an international arbitration process with ICSID
- *Ability of the project company to access justice*, in particular enforcing the government's obligations. After a change of government in Romania in 2000, new conditions, different from the original contract, were imposed on the master commercial concession at Bucharest Henri Coandă International Airport. After the operator (EDF) was unsuccessful on appeal, the government expropriated the business
- *The right or the power (vires) of the contracting authority* (or other key public parties) to agree the project obligations (e.g. does the contracting authority have the right to give the project company the right to operate the airport?) and what administrative or legal steps must be taken to make those obligations binding (e.g. must the PPP agreement be approved by the Ministry of Transport or other regulator before it becomes binding?).

Political risks pose significant challenges for airport PPPs globally. In particular, airport PPPs can be targets for political agendas, where airport licenses are withdrawn, or PPP agreements cancelled following elections or other changes of government.

10.2.2 Legal and regulatory risk

Certain critical legal issues need to be addressed as a prerequisite to implementing PPP. These issues include:[8]

---

7 A sovereign state holds the power of disposition over its territory as a consequence of title (see Brownlie, *Principles of Public International Law* (1990; p. 123).
8 Legal and regulatory risk represents the application of political risks and decisions, discussed further in section 7.1. The close relationship between these sections results in some overlap in the discussions here.

- Authority of the contracting authority to undertake the project. In St Petersburg, Russia, the control of the airport was shared amongst the federal and local governments. The federal government leased its interests to the local government to allow a single contracting authority with the right to implement the entire project
- Procurement rules that permit PPP arrangements
- Security rights over assets and/or shares sufficient to provide the lenders with enough protection
- Protection of the project company against general changes in law and/ or changes targeted specifically at the airport industry/ concessionaire. Most PPP agreements provide for contractual liabilities of the contracting authority for change in law; and
- Access to justice (ideally international arbitration) and a reasonable mechanism for and history of enforcement of judgments/arbitral awards against the government (see Section 11.5 for further discussion on dispute resolution).

A host of other legal issues will be important to ensure the proper functioning of PPP, for example land acquisition, labour relations, tax and accounting (e.g. transfer costs, depreciation, VAT offsetting) and regulatory mechanisms.[9]

10.2.3 Completion risk

The nature of PPP is such that an incomplete project will be of limited value, in particular where the project relates to a new airport (a greenfield investment). PPP allows the contracting authority to package completion risks in a more efficient manner, often known as single point risk allocation. This means that design, construction, installation, commissioning, operation maintenance and refurbishment risk are all allocated to and managed by one entity. Single point responsibility reduces the interfaces between different project functions that can result in errors, delays and a "claims culture" (where different contractors blame each other for any defects discovered – the number of interfaces facilitates such blame games). Under single point responsibility these interfaces are managed by the project company (who is likely more capable of performing this function than the contracting authority).

Completion risk includes:

- The adequacy of the design of the works, including adoption of international or local standards (like ASTM[10]) and consistent with ICAO SARPs[11]

---

9 A full discussion of these issues can be found in Chapter 8 of Delmon, *Private Sector Investment in Infrastructure* (2009); and in Delmon and Rigby Delmon (eds), *International Project Finance and PPPs* (2013).
10 Formerly known as the American Society for Testing and Materials, ASTM International Standards Worldwide is currently one of the standards for contracts around the world.
11 ICAO's Standards and Recommended Practices, see Chapter 5.

200 PPP agreements

- The nature of the technology to be used and the availability of equipment and materials, including transportation, import restrictions, pricing, services necessary for construction, financing costs and administrative costs
- Unforeseen events or conditions, such as weather or subsurface conditions
- The availability of labour and materials, whether skilled labour can be procured locally, to what extent both labour and materials will need to be imported, visas and licenses for such importation and restrictions imposed by local labour laws (including working hours and holiday entitlement)
- The availability of associated infrastructure and services, such as access (road and rail links), water and electricity; and
- The program for completion, whether the construction methodologies are appropriate given seasonal climate, the approvals process, coordination amongst sub-contractors, operations in the airport concurrent with construction activities, and testing and commissioning programs
    - Examples abound of the failure to allocate completion risk, but one case in East Africa saw the contractor build the terminal too close to the runway, rendering the apron in front of the terminal largely unusable. Yet, errors in the allocation of completion risk resulted in the government footing the bill.

10.2.4 Performance risk

In order for the project to maintain sufficient revenues to satisfy debt servicing and to provide a return for the shareholders, the project must deliver infrastructure services to specified levels. Performance risk results in the inability of the facility to deliver the services in the manner and timing required and agreed, in particular failure to comply with agreed levels of service, for example:

- Errors in the design of the facility
- Environmental issues that impede the operation of the facility
- Errors during construction (e.g. safety and security standards); and
- Improper operation of the facility.

Therefore, performance requirements are imposed on the project company by the contracting authority whose requirements are then passed on to the project participants (in particular, the construction contractor and the operator). Whether these requirements are fulfilled by the completed works will be verified by performance tests, as part of the construction regime. During operation the facility will be tested periodically to ensure service delivery. The project company may also wish to obtain further guarantees from suppliers and designers where relevant. These other entities may be best able to cure certain defects or to update technology, as necessary.

Well after the anticipated opening of Bangkok Suvarnabhumi International Airport in 2006, the airport continued to suffer from severe pavement issues including soft spots at the apron and runway/taxiway system. The issues were reported formally by IATA in 2016, suggesting the use of sub-standard construction materials.[12] The airport continued to faces numerous baggage handling issues, as late as six years after opening.

10.2.5 Operation risk

The project must operate to given performance levels in order for the project company to earn the revenues needed to pay operating costs, repay debt and achieve the levels of profit needed. In addition to ensuring that the facility has the performance capacity to achieve these performance levels (as discussed in Section 10.2.4 above), the project company will be required to operate the project in a proper and careful manner, so as to comply with applicable law, permits and consents; and to avoid damage to the project, the site, local or related infrastructure facilities and neighbouring properties. Operation risk will include:

- The risk of defects in design equipment or materials
- The availability of labour and materials
- Changes in operating requirements, owing to changes in law, regulations or other circumstances
- Proper maintenance and the cost of asset replacement and major maintenance.

After the opening of the new Terminal 5 at Heathrow Airport, exclusively dedicated to British Airways, a major operational disruption originated in the baggage handling system, which affected traffic for as long as ten days after opening.

10.2.6 Financing risk

Financing risk relates to the sources of financing to be accessed for the project, the nature of lenders and borrowers, and the constraints imposed by the financial markets at the time of financial close and during the life of the project. This risk can result in increases in the cost of financing and will have a fundamental influence on the financial viability of the project. For example, PPP projects are sensitive to:

- Sufficient tenor of debt (projects with large upfront investment in long lifecycle assets usually look for twelve- to twenty-year debt) and the availability of take-over or refinancing for short tenor debt
- The ability to roll up interest (i.e. pay it later) during a grace period sufficient to address any lack of revenue during the construction period
- Interest rates: the project company will look for fixed rate debt given the fixed nature of some key revenue streams. If fixed rate debt is not available, increases in interest rates beyond that manageable through the revenue stream will need to be hedged or otherwise managed

---

12 "There are also safety concerns about the airport's tarmac, taxiways and apron area because of soft spots," said IATA director-general and chief executive Tony Tyler (*The Strait Times*, February 20, 2016).

202 PPP agreements

- Foreign exchange rates (where the currency of revenues and debt are different – the risk of movements between these)
- The cost of hedging (where interest rate, foreign exchange or other risks are managed through hedging arrangements; the cost and availability of such hedging instruments)
- The availability of working capital financing to cover short-term financing needs; and
- The credit risk of key project participants, including local airlines which may form an important part of demand.

The cost of financing is likely to be uncertain, to some extent, until financial close, since the project company is unlikely to implement any of these financing mechanisms or financial instruments before then. Therefore, the contracting authority will generally share the risk of changes in the cost of financing between bid date and financial close.[13]

10.2.7 Currency risk

Monetary regulation and market conditions can limit the extent to which local currency (capital, interest, principal, profits, royalties, or other monetary benefits) can be converted to foreign currency, how much foreign currency is available and the extent to which local and foreign currency can be transferred out of the country. These restrictions cause significant problems for foreign investors and lenders who will want to have access to distributions and debt service in foreign currencies and to service their debt abroad. As a mainly regulatory risk, this risk is often managed by the contracting authority in developing countries, for example by obtaining waivers or pre-approval for the project company for transfer of foreign currency. Repatriation can be a critical challenge in some countries, for example airlines in Venezuela have had significant challenges repatriating profits.

10.2.8 Demand risk

Demand risk involves any reduction in demand for the services provided by the airport, e.g. fewer passengers, fewer airlines, less cargo.

Forecasts of demand, cost and regulation of the airport sector will be important, including:

- A review of the demand projections and profile, including aeronautical and commercial aspects
- Analyses of prospects for growth, demographic movements, fee structures and airline strategies.

---

13 During this period, there should be a degree of alignment between the interests of the contracting authority and the private partner to achieve financial close promptly. Any change in the base interest rate between bid date and financial close is generally born by the contracting authority, though often limited by a cap or collar.

Where these assessments result in specific risk concerns, the project company may need other support, such as:

- Supplemental revenues from government, e.g. capital grants or availability payments
- Revenue guarantees to ensure a minimum level of revenue
- The ability to adjust the contract performance requirements or contract duration, to compensate for demand shortfalls
- Demand guarantees, to protect the project company from the impact of e.g. traffic lower than forecast/assumed at bid; and/or
- Partial risk guarantee, to protect the project company from shortfalls in the revenue stream associated with specific project risks.

10.2.9 *Environmental and social laws and regulations*

Environmental and social laws and regulations will impose liabilities and constraints on a project. The cost of compliance can be significant and will need to be allocated between the project company and the contracting authority. Equally, in order to attract international lenders, in particular IFIs, the project must meet minimum environmental and social requirements that may exceed those set out in applicable laws and regulations (see Box 10.1). This process is made easier where local law supports similar levels of compliance.

Airport projects generally have a substantial impact on local communities and quality of life, in particular where land must be acquired for airport facilities. Urban encroachment on airport sites is a serious risk in many developing countries, in particular in Africa, creating additional social friction around resettlement of local communities when airports are expanded. Project impact on society, consumers and civil society generally, can result in resistance from local interest groups that can delay project implementation, increase the cost of implementation and undermine project viability.[14] This "social risk" should be high on a lender's due diligence agenda, though it often is not. The lenders and the project company often look to the contracting authority to manage this risk.

Box 10.1 The Equator Principles

The Equator Principles (see www.equator-principles.com) constitute a voluntary code of conduct originally developed by the International Finance Corporation (IFC) and a core group of commercial banks, but now recognised by most of the international commercial banks active in project finance. These banks have agreed not to lend to projects that do not comply with the Equator Principles; these follow the IFC system of

---

14 Delmon, "Implementing Social Policy into Contracts for the Provision of Utility Services" (2007).

> categorising projects, identifying those that are more sensitive to environmental or social impact, and requiring specialist assessment where appropriate. During project implementation, the borrower must prepare and comply with an environmental management plan (EMP).

10.2.10 Force majeure

Certain events, beyond the control of the parties, may inhibit them from fulfilling their duties and obligations under the project agreements. To avoid the resulting breach of contract, the parties will prefer to exclude contractual obligations which have been thus inhibited. Theories of law have developed in response to this need, including the doctrines of impossibility and frustration in English and US law[15] and *force majeure* in French law,[16] where *force majeure* is an event that is unforeseeable, unavoidable and external and that makes execution impossible.[17]

Similarly, the legal systems applicable to the project documents will often provide for an allocation of risk in similar situations, usually suspending all performance requirements, where performance is prevented by such an event. However, each legal system will define *force majeure* events in a different fashion. In order to avoid the potential vagaries and uncertainties as well as the delays involved under applicable law, PPP agreements often provide for a specific regime for *force majeure*, along with a definition of which events shall qualify for special treatment. The term *force majeure* used in drafting project documents comes originally from the *Code Napoléon* of France, but should not be confused with the French doctrine. Generally, *force majeure* means what the contract says it means.

The risk of *force majeure* is generally allocated to the contracting agency to the extent it cannot be insured. The theory goes that the contracting agency is best able to manage *force majeure* risk, as such risk relates partially to the activities of the host country government and its relations with other countries and/or its populace, and that the contracting agency is the only party able to bear such risk, given its size and the difficulty of obtaining adequate insurance. However, in certain markets, the contracting agency may require the project company to bear a portion, or all, of the *force majeure* risk.

The parties will need to discuss how a *force majeure* event will be treated in the context of the project. The *force majeure* regime may include issues of release from project obligations during the duration of the *force majeure* event, payment

---

15 May (6th edition 1995) supra note 224 at 143–150.
16 Liet-Veaux and Thuillier, *Droit de la Construction* (1991, p. 270).
17 "Impossibilité absolute de remplir ses obligations due à un événement imprévisible, irrésistible et extérieur" *French Civil Code*, arts 1147 and 11248 (30 August 1816, reprinted 1991).

by the contracting agency or the offtake purchaser during such period, extension of the concession period, or release from penalties otherwise imposed by the contracting agency for failure to perform in accordance with the requirements of the concession agreement.

The parties may also wish to provide for termination in case of extended *force majeure* events. They may wish to identify a maximum period during which one single event or an aggregate duration of *force majeure* events over the period of the concession may last before one or both of the parties can act to either remove itself from the project or obtain compensation for damages incurred. This period is often linked to insurance cover limitations.

The definition of *force majeure* will vary from project to project and in relation to the country in which the project is to be located. The definition generally includes "risks beyond the reasonable control of a party, incurred not as a product or result of the negligence of the afflicted party, which have a materially adverse effect on the ability of such party to perform its obligations".[18]

In the *force majeure* provision, the parties will generally provide a list, which may or may not be exhaustive, of examples of *force majeure* events. *Force majeure* events generally can be divided into two basic groups, natural events and political events.

Natural events

Although these events form an important part of the *force majeure* concept, they are not automatically attributable or allocable to the contracting agency. The parties will need to look at the availability and cost of insurance, the likelihood of the occurrence of such events and any mitigation measures which can be undertaken. For example, although the contracting agency will be best placed to appreciate the ramifications of national weather patterns and common natural disasters, the project company should be prepared for foreseeable events and should be able to obtain insurance for the majority of this risk.

Natural *force majeure* events may include:

- Unusually severe weather conditions
- Fire
- Adverse natural phenomena, including, but not limited to, lightning, subsidence, mudslide, landslip, heave, collapse, earthquake, hurricane, tornado, typhoon, storm, flood, drought, unusual accumulation of snow or ice, meteorites, volcanic eruption and tidal waves
- Plague
- Acts of God (based on either a definition of this concept in the contract or under the applicable law).

---

18 *Ibid.*

206 PPP agreements

Political and special events

The contracting agency's willingness to protect the project company from political risk will go a long way to reassure the project company and the lenders that the project has host government support. In many developing countries, the risk of political upheaval or interference is of great concern. As a general proposition, the contracting agency in a developing country should be willing to bear a certain amount of political *force majeure* risk. Special risks included in this list generally represent those risks which are not insurable under normal commercial conditions, such as nuclear contamination. These risks are generally considered to be beyond the control of the project company.

The host government may be in such a dominant negotiating position, that it can require the project company to bear some or all of the political project risk. Further, political risk insurances may be available, either through private insurances, multilateral organisations or export credit agencies.[19] Political *force majeure* events may include:

- Terrorism
- Riots or civil disturbances
- War, whether declared or not
- Hostilities, including, but not limited to, sabotage, vandalism and riot
- Blockade or embargo
- Strikes (usually excluding strikes which are specific to the site or the project company or any of its subcontractors)
- Change of law or regulation which materially affects the project
- Nuclear or chemical contamination
- Failure of public infrastructure
- Pressure waves from devices travelling at supersonic speeds.

10.2.11 Default

The project documents will need to co-ordinate termination, the right to terminate, events of default, consequences of termination and step-in rights.

Termination

To avoid the need to refer to national legal systems in order to terminate a given project document, the parties will define situations which may result in termination, and will specify the consequences of such termination by contract, in the project documents and in direct agreements.

Each of the project documents will have its own particular termination regime, with events of default defined in accordance with the obligations of each party. Other default events will be common to the parties such as liquidation or bankruptcy of the other party or extended *force majeure*. However, the

---

19 See Section 5.8.

project documents will need to be co-ordinated to follow closely the termination provisions found in the concession agreement. The project company will usually be able to terminate the PPP agreement for:

- Fundamental breach of the PPP agreement by the contracting agency
- Failure to provide access and possession of the site within a certain period from the effective date
- Failure to provide specified permits and licences
- Extended *force majeure*
- Extended suspension
- Extended failure to make payments due
- Bankruptcy of the contracting agency
- The contracting agency being unable to continue with the project.

The following is a list of events that allow the contracting agency to terminate, which might be found in a PPP agreement:

- Breach of a material provision of the concession agreement
- Failure to commence the works within a certain period from the effective date of the concession agreement
- Failure to achieve completion within a certain period
- Abandonment of a portion of the project without consent
- Bankruptcy
- Change in shareholding of the project company without required approvals
- Replacement of the operator other than in accordance with the PPP agreement
- Replacement of the construction contractor other than in accordance with the PPP agreement
- Calling of loans above a certain threshold related to the project
- Failure to achieve a certain level of performance over time
- A fundamental breach of any of the project documents.

Each project document will need to include a procedure for termination, including notification requirements and remedy periods before a party is entitled to terminate, and set out the consequences of termination, such as compensation, buy-out of the project and other damages. Consistency between the procedures used in the project documents will help manage termination between the various documents. Such provisions will need to take into consideration notice to the lenders and the lenders' right to step in. The contracting agency may also want the right to step in to or buy out the project.

Step-in and continuous operation provisions

The lenders are interested in the success of the project throughout the term of their financing. Where the project company breaches the PPP agreement in such a way as to permit the contracting agency to terminate the PPP agreement, the lenders will want the right of step-in, to continue the project and avoid the termination of the concession agreement. The contracting agency will therefore grant the lenders step-in rights.[20]

In the same way, the project is only of use to the contracting agency when it is in operation. The contracting agency may therefore want the right to continue operation of the project where it terminates the PPP agreement. This is often called the right to continuous operation, as it allows the contracting agency to ensure continuous operation of the project even in the event of termination.

The contracting agency may also wish to have a right to step in to operate the project where, during operation, the project company is temporarily unable to continue operation. Given the urgent need of the host country for the project, the contracting agency may need the right to step in during the temporary disability and ensure continuous operation. Generally:

- Temporary step-in should not involve any reversion of risk to the contracting agency or the offtake purchaser except in cases of negligence; and
- Neither party should gain an additional benefit from temporary step-in; e.g. the contracting agency should be compensated for its reasonable costs of operation.

The lenders and/or the contracting agency may also want step-in rights in relation to the other project documents to ensure the continuity of the project structure and continuity of operation of the project. Thus, each project document, either in the project document itself[21] or in a direct agreement with the lenders and/or the contracting agency, is likely to provide for step-in rights. The project participants may want to receive guarantees from the party stepping in at least as good as those given by the project company.

Transfer to contracting agency

The PPP agreement will need to include a comprehensive strategy for the transfer of the project and its assets to the contracting agency at the end of the period (this procedure may also apply, to some extent, on the termination of the PPP). At the time of transfer the contracting agency may provide the project company with compensation for the residual value of the project assets. This retention of equity return for the end of the project motivates the shareholders to ensure

---

20 See Section 2.8.7 for a further discussion of step-in rights.
21 Provision of the step-in rights in a project document is only effective under legal systems which do not require privity of contract before entitling a non-signatory to rights and obligations under a contract. England is an example of a jurisdiction which requires privity of contract but allows non-parties to be beneficiaries under contracting agency under the Contracts (Rights of Third Parties) Act 2000.

proper operation and maintenance in order to maintain the residual value of the project through to the end of the concession period.

Issues to be considered in establishing a transfer regime are:

- Maintenance schedule prior to the transfer
- Withholding or reserves against completion of outstanding maintenance works during the last twelve to twenty-four months of the project
- Supervision by the contracting agency of the final phase of operation
- Training of contracting agency personnel
- Scope of the transfer, including project assets, intellectual property rights, spare parts, warranties, insurances and other know-how
- Timing of the transfer, including any conditions which must be satisfied at the end of the concession period and any opportunity for extension of the concession period
- Passing of risk to the contracting agency
- Termination of project documents
- Testing before the transfer, including operation, environmental impact and life-cycle forecasts
- Guarantees provided by the project company to the contracting agency in respect of the status of the works
- Termination of all remaining project bonds and guarantees
- Cost of transfer, taxes, fees, repatriation, testing, inspections, and allocation thereof.

10.3 Contractual mechanisms

The risks discussed above will be managed through a series of project agreements between the contracting authority and the project company, and then between the project company and various other project parties.

10.3.1 PPP agreement

Under the PPP agreement (also known as the "concession agreement" or the "implementation agreement"), the contracting authority grants a concession (a series of rights) to the project company to build and operate airport infrastructure for a predetermined period, the "concession period". The PPP agreement may also set out the legal and tax regimes applicable to the project, including the environmental obligations of the project company.

The contracting authority needs to have the legal right to enter into the PPP agreement, i.e. the contracting authority's acts must be *intra vires*. Acts which are *ultra vires* (beyond the power of the party performing the act) may be unenforceable or subsequently rescinded or invalidated under applicable law, i.e. if the contracting authority did not have the right to sign the contract, the project company may have trouble enforcing the contract (it might even be deemed never to have existed).

The primary issues addressed under the PPP agreement will generally be as follows:

- *Completion date*: the contracting authority's need for the infrastructure in question is generally immediate/as soon as possible (often as much for political as practical reasons)
- *Condition of assets*: where existing assets are transferred to the project company, the condition of those assets should be described, with a mechanism to address the possibility that the assets are not actually in the condition anticipated
- *Performance of the project*: the contracting authority's requirements will cover issues such as service standards (generally defined as the level of service (LoS)), efficiency of operation, maintenance needs and costs, lifecycle, health, safety, environmental, quality/quantity of the output/service generated and cost of operation
- *Maintenance regime*: the contracting authority will want to ensure that the maintenance regime (including replacement of parts and materials) implemented is sufficient. This is even more important late in the project; the incentive for the project company to invest funds in maintenance during the final phase of the concession period may be diminished, owing to the imminent transfer of the project to the contracting authority. The contract may offer a withholding of revenues, performance guarantee, or final bonus payment pending quality of the facility to incentivise appropriate maintenance for the entirety of the life of the contract
- *Construction and operation*: the contracting authority will want to ensure that the project company's construction and operation activities meet certain minimum standards (e.g. quality, health and safety), those imposed by law and those specified by the contracting authority to ensure the quality of the services provided and the protection of the public, including timeliness of construction and quality of service rendered
- *Government guarantees*: the government may provide guarantees for public sector bodies taking part in the project whose credit risk is otherwise insufficient; these may be set separately or in the PPP agreement where the government is a party
- *Exclusivity*: the contracting authority may supply the project company with some form of exclusivity rights (see Section 10.1.2, above) over the service to be provided to ensure a bankable revenue stream with careful consideration of future requirements, such as demographic changes. For example, the contracting authority may agree not to develop another airport within a specified distance from the project
- *Know-how transfer*: the contracting authority may want to maximise the interaction between the project company and local personnel in order to ensure the proper transfer of know-how.[22] This is normally ensured through

---

22 For further discussion of the transfer of technology and know-how, see UNIDO's *Guidelines for Infrastructure Development through Build-Operate-Transfer* (1996), pp. 75–90.

the obligation of the project company to engage with an internationally reputable airport operator, often called a technical service agreement (TSA). A typical TSA should consist of the commitment to place foreign managers in executive managerial roles for a considerable period of time, advisory services from abroad, training of personnel at their own airport and the transfer of processes, procedures and systems
- *Government interference*: the contracting authority may undertake that the host government will not act against the interests of the lenders, the shareholders, the project company, the performance of the project company's obligations or the project itself
- *Concession fees*: the project company may be required to pay concession fees for the privilege of obtaining a PPP and to offset contracting authority costs, payable before commencement and possibly periodically during the concession period, e.g. as an upfront fee and/or a share of revenues
- *Restrictions on share transfers*: the contracting authority may want to place restrictions on the transfer or change of shareholding in the project company. The contracting authority may want to disallow any transfer (direct or indirect) until a certain point in time after completion of construction (a lock-up period), a right of approval over the identity of any transferee, and/or to maintain some guarantee from the original shareholders
- *Contracting authority step-in/continuous operation*: the contracting authority may want the right to continue operation of the project where it terminates the concession agreement, sometimes referred to as the "right to continuous operation", to ensure continuous delivery of services
- *Hand-back*: at the end of the concession period, the contracting authority will either put the project out for re-tender or it will require the project company to transfer the project assets to the contracting authority or to a replacement project company.

10.3.2 Construction contract

The project company will allocate the task of designing and building the project to a construction contractor. The construction contract will define the responsibilities of the construction contractor and the project company and their relationship during the period of construction.[23]

The construction phase is generally governed by a turnkey construction contract, sometimes also known as a "design and build" or an "EPC" (engineering, procurement and construction) contract. The term "turnkey" suggests that after completion one need only "turn the key" to commence operation of the constructed facility. Where a single turnkey construction contract is not available, the lenders will want the several contracts to work together as a turnkey

---

23 For further discussion of construction contracts generally, see Scriven, Pritchard and Delmon (eds), *A Contractual Guide to Major Construction Projects* (1999).

contract, or will want a completion guarantee from the sponsors to cover any gaps in risk allocation.

The turnkey construction contract places single point responsibility for the design and the construction of the facility on one party, the construction contractor. Single point responsibility simply means that the construction contractor is bound to provide to the project company a completed project in accordance with the contract specifications, and will be held accountable to ensure that the performance and the quality of the works comply with all of the contractual requirements, so that where there is a problem, the project company has a single point of reference.

Other key terms addressed in a construction contract will include the following (see also the discussion of construction risk in Section 10.2.3):

- *Time for completion*: in PPP projects, timely completion is essential, as penalties may apply under the PPP agreements for late completion, and revenues will generally not be sufficient to meet debt servicing obligations until construction is complete. Turnkey construction contracts are conducive to a fixed time for completion since they combine all design and construction tasks under the single point responsibility of one construction contractor
- *Construction price*: turnkey construction contracts generally use a fixed price lump-sum structure, wherein the contractor is paid one lump sum for the design and construction of the works, with limited opportunity for price increases, so providing greater price certainty for the project company and the lenders. The use of a lump-sum price combined with payments on the completion of stages of construction (for example, milestone payments) can result in an increased rate of progress, since the faster the contractor finishes the construction, the faster they get paid[24]
- *Performance risk*: the completed facility must perform to a certain standard to achieve the revenue stream required to satisfy the debt servicing, provide a return on investment and cover any other costs. The construction contractor will be responsible for constructing works capable of attaining the performance levels required under standard operating conditions, taking into consideration the site conditions, and any other project-specific limitations. Performance tests will verify, for example, lighting, accessibility, ventilation, temperature and environmental impact
- *Site conditions*: the allocation of site risk will depend on the nature of the parties and their expertise. Site conditions, such as geographical, geological, environmental and hydrological conditions, are difficult to define with great accuracy, even after extensive site investigations. Man-made obstructions or conditions may be dealt with separately when allocating this risk, because they are even more difficult to assess accurately by site investigations

---

24 Wallace, *Construction Contracts* (1986), p. 331.

- *Defects liability*: for a period after completion, the construction contractor will remain responsible for the remedy of defects in the works. The period during which the construction contractor is liable for defects is generally called the "defects liability period" although it may also go under the name of "maintenance period" or "warranty period". Certain jurisdictions establish minimum defect liability periods at law.[25]

10.3.3 Operation and maintenance agreement

After completion of construction of the works, the project company will need to operate and maintain the project during the concession period. This function is essential to protect the revenue stream of the project. The operation function will involve managing the operation of the project (for example, specific airport services, the airside, the landside or the entire airport). In order to allocate the risks involved in operation, the project company may contract with an operator. The operation of the project will require an understanding of the local market; the demands on operation in the site country, such as availability of materials and labour for maintenance and repairs; as well as the importance of relationships with local authorities.

The operator's obligations should mirror those set out in the PPP agreement. The key terms addressed in an operation agreement will include:

- *Performance risk*: the operator will be required to work to a standard of performance generally based on the performance standard or operating requirements set out in the PPP agreement, the operation and maintenance manuals and any other supplier instructions in order to maintain relevant warranties
- *Public face of services*: the operator is in a very sensitive position as the operator of an airport, an important public service in the host country. Therefore, the operator's methods of operation and its relationship with its employees and local communities will be critical to the success of the project.

See also the discussion of operation risk in Section 10.2.5, above.

10.3.4 Lending agreements

Financing arrangements for PPP projects follow a general two-step progression. First, funding is provided by the lenders and the shareholders during the construction phase, which is generally considered the riskiest phase of the financing arrangements. Financing during this first phase will include up-front fees, development costs, design and construction. The lenders will advance financing

---

25 For example, decennial liability under French law, Civil Code Article 1792, which makes contractors strictly liable for ten years for all damage, including that resulting from soil conditions, which renders them improper for their intended use.

progressively during the construction phase; payments are usually linked to milestones and verified by an independent expert acting for the lenders and possibly the contracting authority. During this first phase, the lenders will insist on a careful balance of equity and debt financing and may require recourse beyond project assets, to the shareholders or some other guarantor, to cover the risk of any delays or cost overruns which have not otherwise been transferred to the construction contractor.

The second step in project financing is commissioning of construction followed by operation. Completion of construction includes performance tests to ensure that the project is capable of delivering the necessary service levels. Approval of final completion will release the construction contractor from certain liabilities and will therefore be carefully controlled by the lenders. During operation, once the project has begun to produce output, the debt is serviced solely by the project revenue stream.

The lending agreements will therefore set out protections for the lenders, such as:

- Drawdown schedule and the conditions precedent that must be satisfied before each drawdown, in particular related to completion of construction milestones and aggregate paid-up equity
- Lender rights over warranties from contractors, delay-liquidated damages, performance-liquidated damages, contractor-supplied performance bonds, stand-by equity and debt undertakings and other mechanisms to mitigate construction risk
- Funding and control of reserve accounts where the project company must set aside money for contingencies, in particular to cover a number of months of debt service in the event of revenue shortfall, periodic major maintenance expenses and annual costs like insurance and taxes
- Events of default, such as failure to satisfy ratios (debt service cover ratio, loan life cover ratio, debt:equity, etc.), late payment, defaults under project contracts, making changes in management or project contracts without consent; and
- The right for lenders to stop disbursements to the shareholders, control voting rights and other project company discretions ("reserved discretions") and seize funds in the event that things are not going as well as the lenders would like (e.g. where events of default or potential events of default arise).

Security rights (over different project assets and in favour of creditors) are both "offensive" and "defensive": offensive to the extent the lenders can enforce the security to dispose of assets and repay debt where the project fails; defensive to the extent that security can protect the lenders from actions of unsecured or junior creditors. If comprehensive security rights are not available, the lenders may seek to use ring-fencing covenants in an effort to restrict other liabilities, security over project company shares to allow the lenders to take over control of the company, or the creation of a special "golden share" that provides the lenders

PPP agreements 215

with control in the event of default. Security rights may also allow the lenders to take over the project rather than just sell the project assets, since the value of the project may lie in its operation and not in the resale value of the assets.[26]

The lenders and the contracting authority may enter into direct agreements with the project participants setting out step-in rights,[27] notice requirements, cure periods[28] and other issues intended to maintain the continuity of the project where the project company defaults and/or falls away. A project may not require direct agreements where appropriate provisions can be included in the relevant project document or where some other solution is available.

10.3.5 Intercreditor arrangements

Financing for the project is likely to come from several sources, such as commercial banks, multilateral organisations, international financial organisations, and possibly the capital markets, including different levels and classes of debt and equity. An intercreditor agreement will often be used to address key issues,[29] such as:

- Order of drawdown of funds
- Coordinating maturity of loans
- Order of allocation of debt service payments
- Subordination
- Holding and acting on security rights and exercise of discretions, and
- Voting on decisions, e.g. variations of lending agreements, waiver of requirements, acceleration, enforcement of security, and termination of hedging arrangements.

Security and other rights tend to be managed through trustee arrangements, with one of the lenders or a third party acting as agent for the lender group, holding and acting on security rights.

10.3.6 Insurance arrangements

Although the project participants may each provide insurance for the project, it is generally more efficient for the project company to provide or ensure provision of comprehensive insurance coverage for the entire project. In this way,

---

26 Lender rights to run the project rather than just sell off the assets will require consideration of the applicable legal system and its treatment of security and insolvency. Rights over project company shares may achieve the desired security, but may also involve the lenders taking on project risk.
27 Where one party to a contract is in default, the right (usually of the lenders or the government) to take over that party's position in the contract, in an effort to keep the contract from termination.
28 Where one party to a contract is in default, the right (usually of the lenders or the government) to be notified of the default and to be given the opportunity to cure that default.
29 For more detailed discussion of intercreditor agreements, see Wood, *Project Finance, Securitisations and Subordinated Debt* (2007).

the interfaces between different insurance packages, the coverage provided by different insurance providers and the overlapping of the tasks performed by the various project participants will not result in overlapping insurance or gaps in insurance coverage.

Generally, the following insurances will be required:[30]

- Materials and equipment during transportation to the site, including equipment to be integrated into the works, temporary plant and the construction contractor's equipment, ex works to delivery at the site
- Construction all risk (CAR) or construction and erection all risk (CEAR) insurance will cover all operations and assets on the site during construction
- Professional indemnity (PI) insurance for design faults or other such professional services provided by the construction contractor or its designers
- All risk operational damage insurance, including, in particular, insurance of property damage during operations
- Third party liability insurance for any claim by third parties for the acts or omissions of the project company, and any of the contractors, subcontractors or other persons for whom it may be responsible
- Consequential loss insurance, including delayed start up, advance loss of profit and business interruption insurance
- Mechanical or electrical failure not otherwise covered under the operational policy
- Liability insurance for all vehicles to be used on site, including automobiles, refuellers, tugs and tractors, buses, container loaders, potable water truck, belt loaders, and pushback tugs and tractors, which will often be mandatory under local law
- Workers' compensation/employer's liability insurance; and
- Directors' and officers' liability insurance.

The contracting authority will indicate in the concession the insurances it expects the project company to maintain, and to ensure that:

- There is sufficient coverage obtained and maintained
- The contracting authority is co-insured (not joint insured) and that there is no risk of vitiation[31]
- The insurer waives its subrogation rights;[32] and

---

30 Delmon, *Private Sector Investment in Infrastructure* (2016).
31 Where the project insurance involves several insured parties (with varying interests in the insured risk) under the same insurance policy, and the insurance policy becomes unenforceable (with all of the insured parties losing their coverage) due to a breach by one of the insureds of its obligations under the policy (in particular the obligation to disclose relevant information to the insurer).
32 The right of an insurer to take over the rights in action (i.e. right to sue) of its insured, to recover the amount it paid out to the insured.

- Where insurance payments relate to damage to assets that are part of the project, those monies are used to fix the damage or replace the assets, and are not otherwise captured by the lenders or other creditors.

Required insurance may become too costly or unavailable. The parties will need to agree how to manage risks that become "uninsurable", and how to define this term. For certain risks, and in certain markets, the contracting authority may agree to be the insurer of last resort, effectively stepping in to insure risk in exchange for the payment of the premium last paid when the insurance became "uninsurable" or some other agreed rate. However, the contracting authority will want to be sure that the increased costs are not due to project company failure or actions.

Applicable law may require insurance to be obtained locally, in which case the project company may seek to reinsure those risks internationally in order to obtain additional insurance protection. Local law may limit the extent to which reinsurance can be used. Lenders will likely seek cut-through arrangements with reinsurers, to allow direct payment from reinsurers to the project company and/or to the lenders to avoid the risk of insurers not paying or going insolvent after they received payment from the reinsurers.

10.3.7 Guarantees and credit enhancement

A third party (e.g. the government) may provide some form of credit enhancement to reduce the cost of debt or make investments available. This enhancement may be provided to lenders and/or equity investors, compensating them in certain circumstances, or ensuring that a certain portion of their debt service or equity return will be protected. Rating agencies may be consulted when structuring the project to maximise the credit rating for the project (in particular when bond financing is involved), and credit enhancement can result in much higher credit ratings and therefore lower cost of debt (in particular where the credit enhancement brings debt above investment grade – i.e. Standard & Poors BBB–).

Credit enhancement can include:

- Funding/supporting direct payments or grants
- Escrow arrangements or other revenue capture, for example where airport fees must be collected by public entities, those fee revenues can be captured in escrow arrangements to ensure that the appropriate proportion is paid over to the project company
- Providing financing for the project in the form of loans or equity investment
- Providing guarantees, including guarantees of debt, exchange rates, convertibility of the local currency, fee collection (e.g. from a local flag carrier), the level of fees permitted, the level of demand for services and/or termination compensation, etc.
- Providing an indemnity, e.g. against failure to pay by state entities

- Waiving fees, costs and other payments which would otherwise have to be paid by the project company to a public sector entity (e.g. authorising tax holidays or a waiver of tax liability); and
- Providing capital assets or other direct, in-kind investment.

Credit enhancement providers will usually perform their own due diligence on the project, with associated time and cost implications. Providers of credit enhancement may also look for government or other counter-guarantees or security rights, in particular to mitigate the moral hazard of credit enhancement protecting a defaulting party with the perverse incentive for that party to default. Credit enhancement may therefore involve a counter-guarantee from the party whose obligation is being supported.

10.3.8 Sponsor support

The lenders may want access to non-project assets to protect their interests, where the project does not provide sufficient protection to the lenders. So-called sponsor or shareholder support provides the lenders with a guarantee or undertaking from the shareholders (which may need to be supported by bank guarantee, parent company guarantee or otherwise) giving the lenders further security or comfort that the shareholders are committed to the project. Sponsor support may include stand-by subordinated financing for construction cost overruns, guarantees of borrower warranties (in particular those within the control of shareholders), indemnities against environmental hazards and guarantees of cost of materials. The shareholders, however, may want to benefit from limited liability and limited recourse. They will not want to provide further support or increase their liability for the project.

10.3.9 Shareholding arrangements

The shareholders' agreement governs the relationship between the shareholders within the project company, including, for example, a development agreement for the pre-financial close phase, and joint venture agreement for post-financial close. It may include articles of association or incorporation or whatever constitutional documents exist for the project company as well as shareholder loans, stand-by credit, stand-by equity and other similar documentation. The shareholders' agreement will cover topics such as the allocation of development costs, the scope of business of the project company, conditions precedent to its creation, the issue of new shares, the transfer of shares, the allocation of project costs and the management of the project company including decision-making and voting. Such an agreement will often also include a non-competition clause, ensuring that the shareholders do not enter into activities directly or indirectly in competition with the project company.

11 Implementation, monitoring and evaluation

Houston, Tranquility Base here. The Eagle has landed.
Neil Armstrong (1930–2012), American test pilot and astronaut, commander of the Apollo 11 moon landing mission on July 20, 1969

Following financial close, the real work begins: to prepare for the transfer of the airport operations to the private company/consortium and to monitor operations and interfaces during the life of the contract. A skilled government management team should be established for implementation, ideally including members of the project team, and addressing issues like renegotiation of contracts, refinancing, managing the expiration of the contract and handing over of the airport facility to the government.

Governments need to develop robust contract administration processes, during the construction period and during operation and maintenance, through to the end of the contract period.[1] The diagram below maps out the whole of the airport PPP project preparation process, from selection through to implementation. The second diagram provides more detail on the implementation stage of this process.

11.1 Monitoring implementation

During implementation, the contracting authority will establish implementation plans, set up a team to monitor implementation and possibly access external resources to help the parties ensure compliance with contractual obligations, so that the project is implemented well, levels of service are achieved and the project is financially viable.

11.1.1 Implementation team

The government will need to form a project team, with appropriate skills, focused on the transaction (the key management tasks should not be part time jobs), and familiar with the project contracts. The team will need to include financial, legal and technical specialists with access to external advisers. Most importantly, the team leader must inspire confidence in relevant government officials and the private markets, in particular potential investors. The government needs to allocate sufficient budgeting and funding for the team and its functions. The pursuit of funding should not be permitted to distract the team from its key functions, or from its access to key resources, including expert advisers.

Management of a PPP project is not a classic public sector management function, and therefore those establishing the project team and managing relevant budgets and staffing functions need to appreciate the nature of a PPP project and the demands to be placed on the project team. To this extent, the team leader should be sufficiently senior to enable the team to implement its role and to access support and information from other government departments/agencies, with clear lines for decision-making. The team leader should have the full endorsement of the government reform champion, and should have the power to make decisions on the spot without having to consult extensively.

---

1 www.gihub.org.

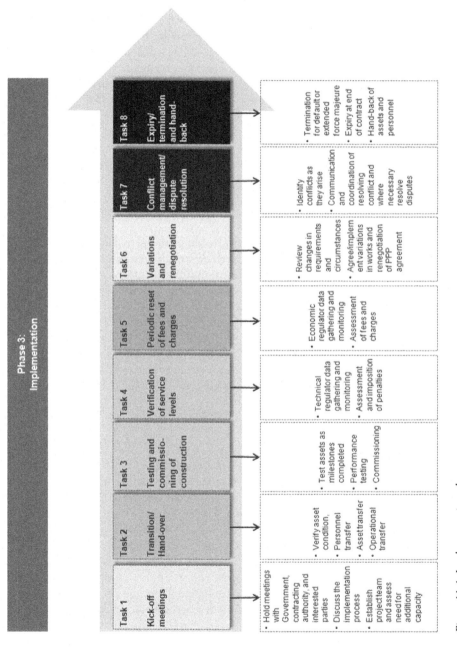

Figure 11.1 Implementation phase

> **Box 11.1 The team and its leader**
>
> The approach of an "Enterprise Team" (ET) has been used by some countries, which is a cross-cutting team with representatives from key stakeholders. The ET is appointed by the Cabinet and is usually headed by someone with private sector experience. In addition, other stakeholders appointed include representatives from the state entity that owns the asset, the PPP Secretariat of the Government and other key ministries, departments and agencies of government for instance, the Line Ministry, the Ministry of Finance, and the Attorney General's Chambers, along with other members from the public and private sector with relevant experience.
>
> The ET has proved to be an efficient way of engaging the key entities that have to be involved in decision making. The decision making authority of the ET members is paramount.
>
> - ET members represent their respective ministries or agencies in the ET meetings where decisions are taken
> - On some occasions however, decisions can be re-opened for discussion if for example the Board of Directors, the government ministers, or the Attorney General expresses a concern or differing views
> - The effectiveness of the ET is heavily dependent on having a strong champion, with sufficient access to decision makers.

"Continuity between personnel involved in the transaction and development and negotiation of concession terms, and mechanisms to institutionalize their knowledge, increases the likelihood of government being able to effectively manage the concession once operational."[2]

11.1.2 Implementation plan

The implementation plan must address the process for transitioning airport operations to the operator and the risks that arise from such process. If the new operator is responsible for airport operations for development throughout the PPP term, the airport transfer takes place after contract finalisation and financial close. In practical terms, the operation may be transferred over a period of time, for example one to three months, during which the new operator becomes familiar with operational details. The PPP agreement may be modified to address any new information that emerges during the transition. For example, in Buenos Aires, companies with commercial contracts signed under the previous administration refused to evacuate the facilities (even when contracts had expired or in cases without any contract), delaying up to one year the time when the new operator could take over the commercial activities. Labour and environmental issues are often discovered only after take-over, requiring further adjustment.

2 IATA, *Airport Ownership and Regulation* (2018b), p. 45.

If the operator assumes responsibilities for the airport operations from day one, and it is determined during operations that further facilities have to be developed, there is a natural incentive for the operator to delay any new investment as much as possible. The longer the operator delays investment, the more of a cash base it can accrue to fund these investments. The financiers may place restrictions on any additional debt of the project company, making additional investments hard to finance. The contracting authority will need to include in the implementation plan a time frame for delivery of any further facilities necessary to avoid delays.

11.1.3 Operation manual

PPP implementation is a complex process, and essential for project success. An "operation manual" maps out the implementation, and can be invaluable for the public and private parties, but is time consuming to develop. The operation manual provides guidance on the implementation process, including:

- A list of required licenses and permits, and key steps in the processes of obtaining them
- All required testing and commissioning regimes, indicating which party is responsible and the framework for supervision and approval
- Sample approval letters, change orders, and other key communications needed for smooth project implementation
- Model process schedules and action plans for e.g. performance assessment
- A summary of rights and obligations under the project agreements, with a time frame for when things must be done, etc.
- A risk matrix and management plan, addressing each risk and how it is to be allocated, monitored, mitigated and managed. Ownership of each risk should be clearly identified on a risk matrix, setting out a clear mitigation strategy; and
- Other practical assistance for project implementation and contract administration.

The manual will help ensure that the rights and obligations under the contract are implemented in accordance with the terms of the contract, keeping track of risks and issues that arise related to the contract.

The operation manual should also include a communications strategy, to ensure regular and ongoing communications on projects and the program, and proactively identify and resolve issues as they arise. Good communication builds trust and enhances the partnership and success of the project and the program.

The government can use the transaction advisers to help it prepare for the role of managing the project, in particular the drafting of the operations manual.

11.1.4 Operational readiness and airport transfer (ORAT)

Opening a new airport, or even a new terminal, may not be as easy as it seems. Even after a well-planned facility is completed, the first day of operation can include numerous surprises. To prevent unfortunate surprises, airports should carry out an ORAT program to allow the existing airport operations to be transferred to the new facility in a seamless manner. The staff may not be properly trained for the new facilities; stakeholders may not be fully prepared; there could be some missing coordination between consulting methodology to ensure the smooth transition from existing operations in the new ones, involving the formulation of processes, training of staff and fool-proofing of new systems and procedures.

There are plenty of infamous examples of new airports or terminals that went through severe problems right after opening. Hong Kong's Chek Lap Kok suffered difficulties of all sorts, including operational, mechanical and organisational, lasting almost half a year. As mentioned in the previous chapter, London Heathrow's Terminal 5 had to cancel over 500 flights during the first two weeks due to major baggage handling system glitches and other problems with car parking; Denver International had major challenges with its baggage system during the first few months after opening. The most dramatic example is probably Berlin Brandenburg airport that was supposed to open by 2011 and to this date has not addressed numerous defects in construction.

11.1.5 Construction, commissioning and performance monitoring

The project company will develop a construction and commissioning plan, approved by the contracting authority. The plan will need to be implemented by the project company and overseen by the contracting authority. Issues to be monitored include those related to design, construction, commissioning and defects liability and other issues that arise during construction.

During the construction stage, the contracting authority will need to monitor and verify that:

- Review and approval of project design has been completed
- All of the land that the project requires for construction to begin has been provided
- Permits and licenses needed for construction have been obtained in a timely manner (environmental permits, zoning permits, building permits, import approvals, etc.)
- All utilities infrastructure, such as approach roads to the airport or electricity, water, and sewerage interconnections, have been completed on schedule
- The performance bond has been received
- All testing during construction, including testing of materials before delivery and installation of the works and associated facilities have been reviewed; and
- Each completion test for each milestone, commissioning of each phase of the works, commissioning of the whole of the works, and performance testing has been reviewed and approved.

During operation of the facility, the parties will need to comply with their obligations including meeting performance obligations and delivering services to standards. Coordination during operation is essential to ensure efficiency and address issues including conflicts as they arise.

During the operational phase of a PPP, the contracting authority will need to monitor and verify:

- Compliance with the project company's obligation to maintain the assets and implement major maintenance programs
- Compliance with performance requirements (based on MTR[3]), including capacity levels achieved; and
- Based on the actual traffic, maintenance of levels of service with minimal disruptions (with some tolerance for short-falls during construction phases).

The contracting authority needs to prepare for the costs and physical requirements of gathering, analysing, and verifying performance data. Contracting authorities not familiar with managing PPP project data will require a steep learning curve.

The contract monitoring team will need to be staffed and funded in a manner sufficient to perform its key tasks and may be supported by an independent monitoring body. The contract monitoring team should:

- Be headed by staff sufficiently senior to elevate concerns quickly and with the mandate to address challenges as they arise
- Avoid conflicts of interest and regulatory capture (where the regulator is captured or effectively controlled by the regulated entities due to low institutional strength of the contract monitoring team, resistant over time due to the long-term relationship between regulator and regulated, etc.)
- Be responsible for all key contractual issues, to ensure coordination, for example verifying service levels, application of fees and charges, and reporting.

Box 11.2 Contract monitoring units: The good, the bad and the ugly

Peru's OSITRAN provides a great example of an effective monitoring body under an independent approach. This Transport Infrastructure Regulatory Agency oversees concession contracts on airports, roads, ports and waterways, and railways. Being a multisector regulator allows the agency more consistent decision-making, ensuring that their impacts on other transport infrastructure are well mitigated, increasing the predictability and clarity for concessionaires and financiers. In practice, specialisation is achieved through smaller multidisciplinary teams per sector, with rotation to other modes to ensure that lessons and best practices are

---

3 Minimum Technical Requirements. Please see Chapter 6 on key performance indicators.

> uniform, with all of them reporting to the supervision and monitoring management.[4]
>
> In Argentina, 35 of the 58 main airports were given in concession to one single operator; a monitoring body named ORSNA (Organism for Supervision of the National Airport System) was created to oversee this. ORSNA has faced numerous challenges. Its design lacks transparency and has been largely captured by the regulated airport company. The airport company breached the contract by not complying with the mandatory works and failing to pay the concession fees from the first day. The contract was later almost completely renegotiated, on terms significantly at the disadvantage of the state. ORSNA was later vested with the authority to carry out investments on its own, managing large sums of monies, despite its institutional shortcomings.

11.1.6 Independent monitoring

The government may be assisted in its monitoring/management function by third parties. For example, an independent specialist may be appointed under the contract to act as a compliance monitor, with specific obligations or assessing investment plans.[5] This independent monitor provides the contracting authority with an independent assessment of performance under the contract.

The independent monitor may also be appointed by the contracting authority and the project company to provide an independent assessment of both parties' performance under the contract. This third-party expert will assess compliance with the contract and also help resolve conflicts that may arise between the parties in relation to contractual obligations.

An independent monitor is often appointed for a limited period of time by small states that do not have the internal capacity to carry out an ongoing monitoring task. For example, a consulting firm could be appointed to oversee the compliance and quality of required investments, or to review and adjust the fees and charges based on predefined criteria.

11.2 Financing

During implementation, the project company will actively manage its financing, to achieve the best blend of equity and debt, to obtain the right source of debt and to reduce the cost of financing where possible. The contracting authority can play an important part in enabling efficient financial engineering, and achieving the best financial solutions for the airport over time.

---

4 Lincoln Flor, Senior Transport Economist, World Bank (Former Manager for Regulation of OSITRAN).
5 Tremolet, Shukla and Venton, *Contracting Out Utility Regulatory Functions* (2004).

11.2.1 Financial close

The signing of a PPP contract does not necessarily lead immediately to the beginning of construction. The project company will likely receive some (20–50%) of its total investment needs from its private owners as equity, and it will want to raise the majority of its finance (50–80%) as debt from lenders. However, the lenders will insist on performing their own detailed due diligence on the project before providing this level of debt. This process can take three to six months to complete (and up to a year for larger, complex projects). The financiers may start due diligence before project award. If they complete due diligence before bidding and the offer of finance is firm, then the bid is "underwritten". Further assessment by the lenders may be required after award but before the offer of financing becomes firm. This involves a second event called "financial close", at which point all conditions precedent have been satisfied or waived and first drawdown of debt can be made.

Reaching financial close can be a very demanding process. Financial close is part of the relationship between private investors and their lenders, however governments or contracting authorities may need to be involved, for example where the lenders:

- Require additional information from the public sector
- Wish to propose changes to the project agreements
- Require direct agreements with or other support from the government/contracting authorities.

In order to establish its role in ensuring timely financial close, the contracting authority may want to:

- Give the project company a maximum period to achieve financial close, generally 12 months from contract award
- Ask to be informed about progress of negotiations, with weekly updates and more detailed periodic reporting, e.g. monthly. The contracting authority may also want access to the lenders to assess whether the project company is proceeding diligently towards financial close
- Respond diligently to requests from lenders for additional information
- Stand ready to enable/encourage timely financial close, including where the investors do not have prior experience in raising project financing for airport PPP projects, the project appears too risky for lenders to finance, changes in the consortium members are required or other challenges arise.

While the contracting authority will not want to take over financing responsibility, failure to reach financial closure is a costly experience for all parties, and will merit efforts by the contracting authority to support the process.

> **Box 11.3 Debt competition**
>
> Mobilisation of debt may be achieved through a competition. The bid process could take into account that, once the preferred bidder is selected, a competition could be run amongst potential financiers. The successful bidder, in coordination with the contracting authority, will run a competition to obtain debt at the best terms. The competition is usually based on a common due diligence report and term sheet produced by the contracting authority. The lending consortium able to commit to the term sheet at the least cost provides debt to the project company.

11.2.2 *Refinancing*

After completion of construction, once construction risk in the project has been significantly reduced, the project company will generally look to refinance project debt at a lower cost and on better terms, given the lower risk premium. In developed financial markets, the capital markets are often used as a refinancing tool after completion of the project, since bondholders prefer not to bear project completion risk, but are often able to provide fixed rate debt at a longer tenor and lower margin than commercial banks. In countries with less developed capital markets, pension funds and other institutional investors may provide refinancing, given their long tenor liabilities.

This refinancing process can significantly increase equity return, with the excess debt margin released and the resultant leverage effect. While wanting to incentivise the project company to pursue improved financial engineering, in particular through refinancing, the contracting authority will want to share in the project company's refinancing gains (for example in the form of a 50/50 split), and may or may not want the right to insist on refinancing when desirable. The contracting authority's project management team will need to have the resources and skills available to manage refinancing issues, either inside the team (less common) or hired in through external experts.

Detailed provisions in the PPP contract generally set out a method for determining and sharing the gains from future refinancing – rather than relying on broad principles and full-blown renegotiation of the contract when refinancing takes place. The specific contract drafting needs to address several points including:

- Deciding when the financing is to take place (the contracting authority may be entitled to call for refinancing in certain circumstances)
- Calculating the expected refinancing gain to the project company shareholders (e.g. using net present value of profits)
- Determining the portion of the gain that should be allocated to each party (e.g. a 50/50 split); and
- Deciding how the gains should be shared (e.g. lump sum payment to the contracting authority.

Implementation, monitoring and evaluation 229

The contract will include other details, for example the discount and interest rates to be used in the calculations and the possible impact of a refinancing operation on termination compensation.

Airports often have multiple phases of construction, where new build is implemented as and when traffic reaches specified levels. Busy airports may be in a constant state of build and expansion. Refinancing for such airports will usually depend on the risk profile for the airport, where traffic achieves certain levels or financial indicators exceed established ratios, indicating the project risk is sufficiently low to merit the cost and effort of refinancing.

In other cases, refinancing may be an essential part of project financing, e.g. when the project can only access short- to medium-term debt (say five to seven years) and project revenues are insufficient to repay the debt during this period, the project company may arrange to repay much of the debt principal in a bullet payment at the end of the debt term. This bullet payment will need to be financed. The risk of the inability to finance the bullet payment will need to be managed, for example with stand-by debt or equity from shareholders, the government or a third party like a bank, a multilateral lender like the World Bank. In some markets, like Australia, PPP debt is often or even mostly funded with short-term debt. Generally, sponsors take the refinancing risk, often along-side the contracting authority.[6]

11.2.3 Selling down equity

Many of the key sponsors are not normally long-term equity infrastructure investors, for example the construction companies who build or refurbish airport facilities, or the equity funds that look for investments with short- to medium-term (three to seven year) exit opportunities. Investors will therefore look for the right to sell down their equity positions as soon as possible. The contracting authority will want the shareholders to remain invested until key project risks have been addressed, in particular construction risks. After completion of construction and proven operation (usually one to three years after commissioning), investors are generally permitted to sell-down part of their equity.

Strategic shareholders are those who provide critical skills/inputs to the project company, e.g. the experienced airport operators. The contracting authority may want to ensure that strategic investors retain sufficient financial interests in the success of the project, to align their interests, for a period long enough to ensure that design and construction meet requirements, the transfer of know-how/expertise from the foreign airport operator to the new company, including management, processes, procedures, systems, etc. and to maintain the political objective of an airport managed by a reputable operator with international brand name and recognition. Strategic investors are often subject to lock-in periods (limitations include when they are allowed to sell down their equity, how much they can sell down and to whom) and may be required to sell shares only to

---

6 "Taking the Risk," *Partnerships Bulletin*, p. 20.

230 *Implementation, monitoring and evaluation*

companies with similar skills, capacity and financial stability to ensure that the project continues to benefit accordingly.

11.3 Renegotiation

Airport PPP projects have the characteristics propitious to recurrent renegotiation; they represent long term, complex commercial and financial arrangements, in a heavily regulated sector, subject to significant political sensitivities, vulnerable to changes in circumstances and often grounded in uncertainty (e.g. traffic demand subject to a wide array of factors beyond the control of the operator, lack of information on business data and ground conditions, hidden liabilities on existing assets, labour, etc).[7] Data from Latin America suggests that some 75% of transport PPP contracts and 87% of water and sanitation PPP contracts are renegotiated at some point.[8] Renegotiations are also pervasive in traditional procurement; for example, change in the scope of the project during construction ("scope creep") is often identified as the primary cause of cost overruns.[9]

Renegotiation is often perceived as failure, as a fundamental flaw in the project, or in PPP generally. This perception arises in particular from poorly managed or implemented renegotiation processes, which:

- Result in terms and conditions less advantageous to one of the parties than those bid[10]
- Result in reductions in mandatory investments, resulting in lower service standards and in a deterioration in the levels of safety and security
- Often lack transparency and are particularly vulnerable to corrupt practices; and
- Can create a public and governmental backlash against private sector involvement in other projects or other sectors, reducing the scope of possible tools that the government will have available to it for improving and reforming its infrastructure services.

---

7 Guasch, Benitez, Portabales and Flor, "The Renegotiation of PPP Contracts: An overview of its recent evolution in Latin America" (2014); Gifford, Bolaños and Daito, "Renegotiation of Transportation Public-Private Partnerships: The US experience" (2014); Bitran, Nieto-Parra and Robledo, "Opening The Black Box of Contract Renegotiations: An analysis of road concessions in Chile, Colombia and Peru" (2012); Engel, Fischer and Galetovic, "Public-Private-Partnerships to Revamp U.S. Infrastructure" (2011); Engel, Fischer and Galetovic, "Soft Budgets and Renegotiations in Public Private Partnerships" (2014).
8 Straub, Laffont, and Guasch (2005). PPP database (preliminary figures): as for 2003, 34% of contracts (by investment amounts) in the water sector are classified as distressed and 12% were cancelled; in transport 15% were distressed and 9% cancelled; while in energy 12% were distressed and 3% cancelled. See http://ppp.worldbank.org.
9 Makovsek (2013) provides an overview.
10 Renegotiations have been used by unscrupulous governments and bidders to pervert an open, competitive bid process by changing the terms of the bid once the "preselected" bidder has won, or to change the scope of the project to include activities for which the government has not obtained approval from the relevant authorities.

Renegotiation is a difficult process, but it is typical for long-term arrangements to face change or conflict and need to address new information and circumstances. PPP agreements therefore include regimes designed to anticipate the need for such adjustments and mechanisms to allow the parties to respond to unforeseeable and unanticipated events and changes that are outside normal business risk. PPP projects must therefore be designed to address change and conflict quickly and effectively, and to facilitate renegotiation in a balanced manner in accordance with the risk allocation framework of the project.[11]

Renegotiations often occur shortly after financial close, even before construction, which is a clear indication of a faulty bid process. Renegotiation at this early stage creates a clear danger of diversion of the procurement process, which could undermine the value for money achieved through competitive procurement.[12] Renegotiation later in the project is more likely to result from changes in external factors. In order to maintain transparency and the rule of law, governments should disclose the entire renegotiation process and provide detailed justification for the change in conditions and of the fairness of any changes in the terms of the PPP.

Box 11.4 Renegotiations in practice

Because of the 2001 world crisis affecting the aviation sector, compounded by the financial crisis in Argentina, Santiago de Chile airport lost a significant amount of its traffic. The government renegotiated the PPP agreement switching to a variable term, extending the concession to guarantee a fair return on investment.

In July 2009, with over 60% of the 600m greenfield Quito Airport complete, the Constitutional Court of Ecuador ruled that airport fees and charges were state property, which fundamentally undermined the PPP agreement and financial arrangements for the project. The arrangements needed to be renegotiated quickly to avoid delay of project construction works. The renegotiation lasted 18 months and resulted in a relaunch of construction and successful completion of the project.

The reasons for renegotiation are many, but some of the aggravating factors can be summarised as follows:

- One party is doing better/worse than it hoped – especially when the private sector makes more profit than is politically comfortable, or the project is not sufficiently profitable to hold the investor's attention. In some cases, an opportunistic firm makes an overly aggressive offer and must renegotiate after

---

11 Delmon and Phillips, "Renegotiation of Private Participation in Infrastructure and the World Bank" (2007).
12 Guasch, Benitez, Portabales and Flor (2014); Engel, Fischer and Galetovic (2014).

award to make the deal financially viable. The question of intentionality is a difficult one. Some complications might include:
- An oddly low or aggressive bid
- A contract that leaves openings for change without sufficient limitations or oversight
- The government has in the past allowed such negotiation, or has given indications that it would be open to such calls for negotiation
- An opportunistic contracting authority wants to expand the scope of works without having to obtain approvals or beyond agreed fiscal limits, where permitted or where oversight is insufficient
- Parties lacked sufficient information on the project, e.g. where a full feasibility was not performed, or was performed by a biased party, or where there was lack of time and opportunity during bidding process to obtain and assess data
- Change in circumstances after financial close (e.g. market volatility, cost increases, change in demand patterns, political/legal changes), when the PPP agreements does not contemplate such changes
- Change in contracting authority requirements not included in the PPP agreement (for example, a change in law or regulations that alter security standards).

Developing a framework for renegotiation

Chile has one of the most successful PPP programs among developing countries. Their PPP Act allows the parties to agree to change the works and services contracted, in order to raise the service levels and technical standards by up to 15% (which figure is established in the bidding conditions) of the approved capital value. The Ministry of Public Works must be able to justify the changes for duly substantiated reasons of public interest in a public report. The circumstances giving rise to the amendment must occur after the awarding of the concession, and could have not been foreseen upon its awarding, and awarding the new works to the original concession holder is more efficient than granting a new concession. If the increase exceeds 5% of the approved capital works then it must be put out to open tender by the private partner. There have been 148 renegotiations of PPP contracts; on average each contract had been renegotiated three times.

Box 11.5 Lessons learned from Australian renegotiation

In Australia, under the country's PPP Guidelines, renegotiation of any significant areas of a PPP contract after it has been approved and signed by government will require the contracting authority to obtain Cabinet approval prior to commencing negotiations.

The ability to manage the changing environment by adapting contracts has been key to de-stressing projects. On the whole, the renegotiation outcomes in Australia have been beneficial and demonstrated government commitment to managing PPPs in a manner that is in the public interest. The following are words of wisdom from an ex-official from Australia on the process of renegotiation:

"Australian PPP agreements cater for changes in a variety of ways:

- Empowered committees to apply specifications in a manner that drives efficiency in public facilities such as hospitals
- Dispute resolution procedures permit mediation and facilitated agreements
- Amendments are permitted and subject to the same oversight as the original approvals and gateways.

So far, the market works. The state governments have sent a strong message to the market that they will not guarantee any private sector debt and will not bail out projects. The projects do not go under, the concessionaire is liquidated, and equity is sold into a secondary market in a process led by lenders."

Source: *National Public Private Partnership Guidelines*, December 2014, Volume 6.

Practical guidance on renegotiation

A number of good practices have been developed on renegotiation:

- Feasibility studies can help identify key issues early to avoid the need for renegotiation
- Project contracts need to address issues as completely as possible and to provide mechanisms to address conflict, changes and circumstances. Balanced risk allocation is key. Some contractual clauses are more likely to create dispute, for example financial equilibrium clauses have been shown to complicate renegotiation[13]
- Proactive conflict identification, management and resolution helps keep communication open and address problems as they arise
- Transparency, competition, an efficient approvals process, a robust institutional framework, public disclosure of PPP contracts all help improve communication and avoid renegotiation
- Involvement of independent monitors, like sector regulators to respond to changes, can ensure any renegotiation is balanced and is only permitted if absolutely necessary

---

13 Guasch, Benitez, Portabales and Flor (2014).

- Some countries impose a holding period for renegotiations for the first three years of a project, to avoid the use of renegotiation to compensate for a low-ball bid, for example in Peru and Colombia
- Bid evaluation based on best overall economic bid rather than lowest price reduces the likelihood of renegotiation
- If significant changes are to be made in the scope of works, such work must be tendered competitively, to avoid abuse of the renegotiation process.

Box 11.6 The renegotiation process of San José Juan Santamaría Airport

In 2001, the Costa Rican State granted a PPP for Juan Santamaría International Airport to Alterra Partners. However, in 2003 due to escalating construction costs and delays in the progress of the work, the company requested increases in the fees and charges. Because of this dispute, the company froze all construction. In February 2006, after several years of legal disputes between the Civil Aviation Technical Council (CETAC) and Alterra, the resumption of the construction works was announced by the government after an order from the Comptroller General of the Republic, which concluded that the financial deficit of Alterra Partners was not a valid reason to stop the expansion. However, in December 2006, the government of Costa Rica announced the "anticipatory termination" of the PPP agreement, as the lenders would not ensure necessary disbursements to cover construction, given Alterra's financial problems.

Finally, in June 2009, after almost eight years of conflict, Alterra Partners decided to sell the project to Aeris Holding Costa Rica S.A., with financing from the Inter-American Development Bank and Overseas Private Investment Corporation (OPIC), for the completion of the construction works (to be paid by Aeris).

This has been one of the longer lasting conflicts in a concession that did not end in a cancellation or expropriation.

Sources: "Aeropuerto Juan Santamaría: la puerta de entrada a Costa Rica", *Diario La Nación*, 26 February 2017; "Gobierno anuncia término de concesión de aeropuerto con británica Alterra", *Diario La Nación*, 18 December 2006; "BID presta $45 millones para ampliar Juan Santamaría", *Diario La Nación*, 4 December 2009.

11.4 Step-in

PPP contract terminations are expensive for all sides involved, and that is why they are relatively rare in large, capital-intensive PPPs. Almost everyone finds it more cost-effective to try to restructure and renegotiate these deals than to incur the cost and pain of cancellation of PPP contracts.

Where the project company breaches the PPP agreement in such a way as to permit the contracting authority to terminate, the lenders will want the right of step-in before the contracting authority is allowed to terminate, so that they can cure the breach, continue the project and avoid the termination of the concession agreement. For further discussion of lender step-in see Section 10.3.4.

The contracting authority may also wish to have a right to step in to operate the project where, during operation, the project company is temporarily unable to continue operation. Given the urgent need of the host country for the project, the contracting authority may need the right to step in during the temporary disability and ensure continuous operation. Generally:

- Temporary step-in should not involve any reversion of risk to the contracting agency except in cases of negligence; and
- Neither party should gain an additional benefit from temporary step-in (e.g. the contracting agency should be compensated for its reasonable costs of operation).

The lenders and/or the contracting authority may also want step-in rights in relation to the other project contracts (i.e. the project company's sub-contracts) to ensure the continuity of the project structure and continuity of operation of the project where the project company cannot fulfil its obligations, but the project company's sub-contractors are fulfilling their obligations. For example, if there is a termination of the project during construction, the lenders and/or contracting authority may want a contractual arrangement with the construction contractor to keep any disruption of construction to a minimum. Thus, each key project contract, either in the contract itself or in a direct agreement with the lenders and/or the contracting authority, may provide for step-in rights. The project participants may want to receive guarantees from the party stepping in at least as good as those given by the project company.

To be clear, the step-in regime is a contractual mechanism to address scenarios where the project company is unable or unwilling to fulfil its obligations and other project parties want to maintain the project. This is distinct from the government's right in law to expropriate project assets (usually limited and requiring just compensation), which is distinct from unlawful expropriation, where the government seizes assets without legal entitlement.

11.5 Conflict management and dispute resolution

The long-term nature of PPP, the variety of stakeholders and interested parties, public and private sensitivities and different methodologies for resolving conflict can result in frequent conflict and the potential for disputes. A well-designed PPP will pay special attention to conflict management, identifying conflict early, and elevating it to the right level for resolution. Where conflict management fails, and the parties find themselves in dispute, the resolution of the dispute needs to focus on the project. A failed project is a disaster for all parties; there are seldom true "winners" from a failed project, no matter what might be reported by interest groups or board meetings.

Dispute resolution should be kept in the hands of those who know the project best, i.e. the parties themselves. Non-binding forms of dispute resolution can be used to facilitate a settlement, and in some cases to reinforce the relationship between the parties – resolving difficult issues can cement a relationship and prove commitment to the success of the project.

Where non-binding mechanisms are not successful, access to binding dispute resolution is essential. PPP projects often involve parties from a variety of legal, social and cultural backgrounds. Identifying a national court capable of meeting the needs of such a diverse collection of parties, and acceptable to each of these parties, may be difficult. It is for this reason that parties to a PPP project generally prefer to submit any disputes to arbitration rather than to state courts.

It is important that the dispute resolution regimes in the different project documents are similar, if not identical, in order to facilitate the resolution of complex issues involving multi-party resolution before a single forum.

This section will discuss some of the mechanisms for conflict management and non-binding and binding dispute resolution.

11.5.1 Conflict management

Given the complexity of airport PPP projects and the diversity of parties, conflict management is key to project success. A culture of conflict management should permeate the project documents and the parties' interactions. Communication is critical, keeping dialogue open and constant. Where conflicts arise, they need to be identified and addressed early and proactively.

Projects can benefit from partnering, whereby all participants meet together prior to commencement to establish communications, expectations and responsibilities. The partnering process is usually professionally facilitated and the goal is to establish among all stakeholders and participants a reliable and mutually shared system of accountability for communications.

A clear mechanism of escalation of concerns and conflicts can help notify decision makers of the issues and engage their support in resolution. Of course, no one will want to bother their senior management or chief executive officer for minor issues, but these same small issues can grow to serious problems if not resolved and maybe the support (or threat) of senior management involvement is what is needed to encourage creative solutions.

11.5.2 Early evaluation

Early evaluation involves review of the dispute by an independent expert to seek resolution early and in a short time frame, without the cost of arbitration. A few common, early dispute resolution mechanisms include:

- Parties to a PPP may jointly appoint an independent engineer to oversee construction and operation, monitor testing and play a dispute resolution role
- Some countries mandate early expert assessment for certain disputes, for example "adjudication" for domestic construction contracts in the UK[14] is mandated before reference can be made to final and binding arbitration or the courts
- Many large, complex projects adopt dispute resolution boards ("DRBs"), created at the inception of the project, whose purpose is to monitor the project status and be available for stakeholders in the event of disputes.[15] The DRB members follow the project from commencement to completion, keeping up with issues that arise in relation to the project, the parties involved, and the various project obligations. From the data gathered by the DRB Foundation, dating back to 1975, 60% of projects with a DRB had no disputes, and 98% of disputes referred to a DRB did not result in subsequent litigation or arbitration.[16]

11.5.3 *Mediation*

Mediation is a powerful mechanism for conflict and dispute resolution that is under-utilised in PPP. Mediation is simply a facilitated negotiation, with no decision or factual interpretation.[17] The mediator may be called upon to give a suggested resolution (often called "evaluative" mediation or "conciliation" as compared to the "facilitative" model of mediation), but will not bind the parties.

Mediation is very flexible, allowing the mediator to structure the process in the manner most appropriate to the situation. There are several points at which mediation can be used in PPP projects:

1. *Consultation*: during project preparation, governments need help consulting stakeholders (community groups, labour unions, land owners, government staff, etc.); during these processes there may be opportunities for a mediator to help facilitate dialogue and open processes for sharing of issues while managing stress and frustrations
2. *Negotiation*: during negotiation, there might theoretically be an opportunity for a mediator to help resolve deadlock on specific issues, or more generally to help bridge cultural divides between the parties in order to finalise the deal
3. *Initiation*: at the very beginning of a project, once the ink dries on the contracts and the difficult work of implementation begins, it is useful to bring

---

14 Part II of the Housing Grants, Construction and Regeneration Act 1996.
15 The Dispute Resolution Board Foundation www.drb.org provides further information about DRBs and has branches in many major jurisdictions.
16 www.drb.org.
17 Delmon and Phillips (2007); Mackie, Miles and Marsh, *Commercial Dispute Resolution* (1995); Delmon, *BOO/BOT Projects: A Commercial and Contractual Guide* (2000), p. 248.

the parties together and have a mediator walk them through the practical process of implementation (similar to partnering), setting out or affirming implementation plans, teasing out details of how things will be done in practice, who is responsible for what and when. The goal is to establish among all stakeholders and participants a reliable and mutually shared system of accountability and communications to enhance the team's ability to identify problems at a very early stage, when solutions are cheapest and most plentiful
4. *Deal*: there are a number of changes, conflicts, challenges that arise in a project which do not amount to a "dispute" under the relevant contractual clauses. For example, where the parties need to agree a variation order or there is an opportunity for refinancing, there may be disagreements about timing, currency, or tenor that a mediator could help resolve, but which no one will want to bring to the formal dispute resolution mechanism
5. *Dispute*: where there is a full-blown dispute, whether or not the parties have included mediation in the contract, access to mediation can help avoid costs and delay. It is common for commercial parties to agree, in their initial contracts, to a stepped dispute management process under which they require mediation prior to asserting claims to binding dispute resolution.

11.5.4 Arbitration

Arbitration is most often used as the final method of dispute resolution (instead of litigation) in airport PPP projects because of its flexibility and greater ease of award enforcement.[18] Arbitration is a private dispute resolution process, which produces a binding result, immediately enforceable at law under most legal systems through arbitration acts and treaties. The mechanism by which the tribunal reaches its decision, known as an award, is generally subject to due process and each party must have a reasonable opportunity to present its case. The parties determine the type of arbitration and seek to agree on the appointment of the members of the tribunal. Like other private systems, arbitration is also private in the wider sense in that it is confidential. The parties' agreement to arbitrate may be a clause that forms part of a contrac;, it may be a separate agreement entered into by the parties before or after a dispute has arisen; or it may form part of an international treaty, for example a Bilateral Investment Treaty (BIT). The following addresses a few issues specific to arbitration.

- *Confidentiality*: arbitration is a private process, and the parties can undertake a general duty of confidentiality. This is not normally absolute but is qualified by exceptions such as compulsion of law or where production of documents is necessary to protect the legitimate interests of one of the parties
- *Choice of arbitrator*: arbitration allows the parties to appoint someone with experience of the subject matter of the dispute, rather than entrusting the

---

18 See generally Sutton, Kendall and Gill (eds), *Russell on Arbitration* (1998).

resolution of a technically complex issue to a legally trained, but scientifically inexperienced, judge. In most cases, a panel of three arbitrators are appointed, one appointed by each party and the third (the President of the Tribunal) agreed between the first two arbitrators. Appointing authorities (see below) can also help the parties identify arbitrators, where there is no agreement

- *Enforceability of arbitral awards*: many countries have signed treaties providing for reciprocal enforcement of arbitral awards. These treaties, such as the New York Convention,[19] do not usually allow the enforcing court to open up the award and make a qualitative assessment of its merits except in limited circumstances. The enforcement of an arbitral award may be easier than that of a judgement obtained through the courts because of the wider application of international conventions relating to arbitral awards
- *Arbitration rules*: the parties may stipulate the procedure to be followed by an arbitral tribunal. They may appoint an arbitral institution (e.g. the International Chamber of Commerce (ICC), the London Court of International Arbitration (LCIA), the Arbitration Institute of the Stockholm Chamber of Commerce, or the Singapore International Arbitration Centre) to administer the arbitration and run it according to the relevant institutional rules. UNCITRAL[20] provides a set of rules sometimes viewed as more neutral, which can be implemented by the relevant institution. A separate arbitral body has been developed under the International Centre for Settlement of Investment Disputes (ICSID), part of the World Bank Group, with greater powers of enforcement. A strict procedure must be followed in order to submit disputes arising under a contract to ICSID.[21] Bilateral Investment Treaties (BITs) can provide a method for accessing ICSID arbitration if the same is not contemplated in the contract
- *Arbitration procedure*: the parties may agree on the number of written pleadings and the time limits for their submission. The written procedure usually consists of two rounds of pleadings. During the first round the requesting party files a *Memorial* followed by the opposing party's *Counter-Memorial*. During the second round the requesting party files a *Reply* followed by the opposing party's *Rejoinder*. The Tribunal may also order, on its own accord or as agreed by the parties, the filing of other submissions such as post-hearing briefs. Pleadings generally consist of statements of facts, legal submissions and the request for relief. They may be accompanied by relevant evidence, in particular witness statements, expert reports and exhibits.[22]

---

19 United Nations Convention on the Recognition and Enforcement of Foreign Arbitral Awards (1958).
20 United Nations Commission on International Trade Law.
21 www.icsid.worldbank.org.
22 See for example "Written Procedure: ICSID Convention Arbitration" (ICSID, 2019).

11.5.5 Litigation

Litigation is a public dispute resolution process and is the default dispute resolution system where no other system has been stipulated by the parties, although it can also be expressly chosen. Unlike arbitration, where the parties bear the full cost of the tribunal, litigants pay only nominal fees to the court. However, this cost saving may be outweighed by the costs incurred by the parties in complying with the lengthy procedural requirements of litigation.

In the context of international projects, it is important to bear in mind that litigation is a national dispute resolution process, tied to a particular national system of courts. Problems may arise when the defendant is located in a different country to that of the relevant court. For instance, there may be various specific procedural requirements for commencing an action against a foreign defendant. There may also be a perception of bias against foreign litigants. Even if no actual bias exists, foreign claimants will prefer to submit to a process with which they are comfortable.

A further issue which arises in an international context is the enforcement of judgements. For enforcement of a foreign judgement, each country will have its own rules and will be party to different international conventions. Different procedures must be followed in respect of each of these systems, and there are different substantive grounds for the court to refuse enforcement. The process of enforcement of a foreign judgement is therefore generally more complex and difficult than that of an arbitral award, unless the judgement is subject to arrangements for reciprocal recognition, for example under the Convention on Jurisdiction and the Enforcement of Judgments of 1973 (the Brussels Convention) and the Lugano Convention of 1988 amongst certain European nations.

11.6 Expiry, termination and handover

After delivery, whether the project is terminated early, or expires in accordance with expectations, the parties will need to manage the exit phase, for example to:

- Identify relevant assets and other things that need to be transferred and assess their value. Some assets will transfer automatically, and some only where the contracting authority so chooses
- Assess the condition of assets to be handed over, ensure those assets are handed over in the agreed condition. In the run up to expiry, a regime is normally agreed in the contract to assess the condition of assets and resolve any defects
- Monitor and ensure remedy of any defects or deficiencies in those assets and resolve issues associated with the allocation of any costs of any such remedy
- Procure a replacement project company (or provide access to information to do so), when needed; and
- Monitor the transfer of assets, staff and the business generally to the contracting authority or an appointed entity.

Bibliography

Agénor, P.-R. and Moreno-Dodson, B. (2006). "Public Infrastructure and Growth: New channels and policy implications", World Bank Policy Research Working Paper 4064.

Airports Council International (2018). Media Release: "ACI's World Airport Traffic Forecast reveals emerging and developing economies will drive global growth", 1 November, available online at: https://aci.aero/news/2018/11/01/acis-world-airport-traffic-forecast-reveals-emerging-and-developing-economies-will-drive-global-growth/

Airservices Australia (2010). "Terminal Navigation Pricing Review Discussion Paper", available online at: www.airservicesaustralia.com/services/charges-and-costing/pricing-proposal/2011-pricing-proposal/2010-terminal-navigation-services-pricing-review/

ASEAN (2015). "Building the ASEAN Community: ASEAN Single Aviation Market", available online at: www.asean.org/storage/images/2015/October/outreach-document/Edited%20ASAM-2.pdf

Aviation: Benefits Beyond Borders (2016). "Powering Global Economic Growth, Employment, Trade Links, Tourism and Support for Sustainable Development through Air Transport", available online at: https://aviationbenefits.org/media/149668/abbb2016_full_a4_web.pdf

Bartlett School of Planning (n d). "UCL Project Profile on Greece, Attiki Odos", available online at: www.omegacentre.bartlett.ucl.ac.uk/publications/omega-case-studies/

Benavente, P. (2011). "Rol y funciones del Organismo Regulador", OSITRAN, available online at: www.gub.uy/ministerio-economia-finanzas/sites/ministerio-economia-finanzas/files/documentos/noticias/benavente---rol-y-funciones-de-ositran.pdf

Bitran, E., Nieto-Parra, S. and Robledo, J. (2012). "Opening the Black Box of Contract Renegotiations: An analysis of road concessions in Chile, Colombia and Peru", Draft Renegotiations in Chile, Peru and Colombia, Mimeo.

Boeing (2019). "Boeing Commercial Outlook: Commercial Market Forecasting", May, available online at: www.boeing.com/commercial/market/commercial-market-outlook/

Borges, J.L. (1970). "Preface," Dr Brodie's Report [El informe de Brodie], available online at: https://theteacherscrate.files.wordpress.com/2015/09/borges-jorge-luis-doctor-brodies-report-bantam-1973.pdf

Braude, J.M. (1968). *Braude's Source Book for Speakers and Writers*. Upper Saddle River, NJ: Prentice-Hall.

Brixi, H., Budina, T. and Irwin, N. (2006). "Managing Fiscal Risk in Public Private Partnerships", World Bank; a later version, "Public-Private Partnerships in the New

EU Member States: Managing fiscal risks", available online at: http://siteresources.worldbank.org/INTDEBTDEPT/Resources/468980-1207588563500/4864698-1207588597197/WPS114.pdf.

Brownlie, I. (1990). *Principles of Public International Law* (fourth edition). Oxford: Clarendon Press.

Business Roundtable US (1982). "Contractual Arrangements: A construction industry cost effectiveness project report", October, *The Business Roundtable*, New York.

Calderón, C. and Servén, L. (2010a). "Infrastructure and Economic Development in Sub-Saharan Africa", *Journal of African Economies* 19, AERC Supplement 1, 13–87.

Calderón, C. and Servén, L. (2010b). "Infrastructure in Latin America", Policy Research Working Paper 5317. Washington DC: World Bank.

Calderón, C., Moral-Benito, E. and Servén, L. (2011). "Is Infrastructure Capital Productive? A dynamic heterogeneous approach", Policy Research Working Paper 5682. Washington DC: World Bank.

CAPA (2018). "The Airline Cost Equation: Strategies for competing with LCCs", 14 May, available online at: https://centreforaviation.com/analysis/reports/the-airline-cost-equation-strategies-for-competing-with-lccs-416644

Chong, J. (2017). *Legal and Social Affairs Division: Airport governance reform in Canada and abroad*. Canada: Library of Parliament.

Civil Aviation Authority, UK (2014). "Economic Regulation at Heathrow from April 2014: Final proposals", available online at: https://publicapps.caa.co.uk/docs/33/CAP%201103.pdf

Colman, N. (2001). "The Manly Wisdom of Will Rogers", in *The Friars Club Bible of Jokes, Pokes, Roasts, and Toasts*. New York: Black Dog and Leventhal Publishers.

de Neufville, R. (1999). *Airport Privatization Issues for the United States, Technology and Policy Program*. Cambridge, MA: Massachusetts Institute of Technology.

Delmon, J. (2000). *BOO/BOT Projects: A Commercial and Contractual Guide*. London: Sweet & Maxwell.

Delmon, J. (2005). *Project Finance, BOT Projects and Risk*. The Netherlands: Kluwer Law International.

Delmon, J. (2007). "Implementing Social Policy into Contracts for the Provision of Utility Services", in A. Dani, T. Kessler and E. Sclar (eds), *Making Connections: Putting Social Policy at the Heart of Infrastructure Development*. Washington DC: World Bank.

Delmon, J. (2010). "Understanding Options for Public-Private Partnerships in Infrastructure: Sorting out the forest from the trees: BOT, DBFO, DCMF, concession, lease …", World Bank Working Paper. Washington DC: World Bank.

Delmon, J. (2011). *Public-Private Partnership Projects in Infrastructure: An essential guide for policy makers*. New York: Cambridge University Press.

Delmon, J. (2016). *Private Sector Investment in Infrastructure: Project finance, PPP projects and PPP programs* (third edition). Alphen aan den Rijn, Netherlands: Kluwer International.

Delmon, J. (2017). *Public-Private Partnership Projects in Infrastructure: An essential guide for policy makers* (second edition). New York: Cambridge University Press.

Delmon, J. and Delmon, R. (eds) (2013). *International Project Finance and PPPs: A legal guide to key growth markets* (third edition). Alphen aan den Rijn, Netherlands: Kluwer International.

Delmon, J. and Phillips, R. (2007). *Renegotiation of Private Participation in Infrastructure and the World Bank*. Washington DC: World Bank.

Delmon, V. (2008). "Airport BOTs/Concessions-Checklist of Legal and Regulatory Issues, PPP", in *Infrastructure Resource Center for Contracts, Laws and Regulations*. Washington DC: World Bank.

Bibliography 243

Diario La Nación (2006). "Gobierno anuncia término de concesión de aeropuerto con británica Alterra", Diario La Nación, 18 December.
Diario La Nación (2009). "BID presta $45 millones para ampliar Juan Santamaría", Diario La Nación, 4 December.
Diario La Nación (2017). "Aeropuerto Juan Santamaría: la puerta de entrada a Costa Rica", Diario La Nación, 26 February.
Doganis, R. (2001). *The Airline Business in the 21st Century*. London: Routledge.
Engel, E., Fischer, R. and Galetovic, A. (2011). "Public-Private Partnerships to Revamp U.S. Infrastructure", Hamilton Policy Brief, Brookings Institution.
Engel, E., Fischer, R. and Galetovic, A. (2014). "Soft Budgets and Renegotiations in Public Private Partnerships", Discussion Paper No 2014–17, International Transport Forum at the OECD, Paris.
Engel, E., Fischer, R. and Galetovic, A. (2016). "The Joy of Flying: Efficient airport PPP contracts", available online at: https://extranet.sioe.org/uploads/sioe2016/engel_fischer_galetovic.pdf
Estache, A. and Garsous, G. (2012). "The Impact of Infrastructure on Growth in Developing Countries", *Economics Notes* 1. Washington DC: International Finance Corporation.
Financial Times (2011). "BA buys BMI Heathrow slots", *Financial Times*, 23 September.
French Civil Code (1816, reprinted 1991). "Impossibilité absolute de remplir ses obligations due à un événement imprévisible, irrésistible et extérieur", articles 1147 and 11248.
Gifford, J., Bolaños, L. and Daito, N. (2014). "Renegotiation of Transportation Public-Private Partnerships: The US experience", Discussion Paper No 2014–16, International Transport Forum at the OECD, Paris.
Graham (n d). *Fundamentals for Airport Privatization and Concession Policies*. London: University of Westminster.
Guasch, J.L., Benitez, D., Portabales, I. and Flor, L. (2014). "The Renegotiation of PPP Contracts: An overview of its recent evolution in Latin America", Discussion Paper No 2014–18, International Transport Forum at the OECD, Paris.
Haley, G. (1996). *A-Z of BOOT*. London: IFR Publishing.
Heathrow Airport Holdings Limited (2018). "Annual Report and Financial Statements for the Year Ended 31 December 2018", available online at: www.heathrow.com/file_source/Heathrow/Static/PDF/Heathrow-SP-Limited-Q4-2018-results-release.pdf.
HM Treasury, UK (2006). "PFI: Strengthening long-term partnerships", available online at: http://www.hm-treasury.gov.uk/media/7/F/bud06_pfi_618.pdf
HM Treasury, UK (2018). "Green Book", available online at: https://assets.publishing.service.gov.uk/government/uploads/system/uploads/attachment_data/file/685903/The_Green_Book.pdf
Hodges, J.T. and Dellacha, G. (2007). "Unsolicited Infrastructure Proposals: How some countries introduce competition and transparency, An international experience review", World Bank Working Paper. Washington DC: World Bank, available online at: http://documents.worldbank.org/curated/en/142981468777252745/Unsolicited-infrastructure-proposals-how-some-countries-introduce-competition-and-transparency.
Hussain, M.Z. (ed.) (2010). "Investment in Air Transport Infrastructure: Guidance for developing private participation", World Bank Working Paper. Washington DC: World Bank, available online at: http://documents.worldbank.org/curated/en/228741498036938148/Guidance-for-developing-private-participation
IATA (2017a). "Improved Level of Service Concept", IATA Consulting, available online at: https://www.iata.org/services/consulting/Documents/cons-apcs-los-handout.pdf

IATA (2017b). "Worldwide Slot Guidance" (eighth edition), effective 1 January.
IATA (2018a). "IATA Forecasts $35.5bn Net Profit for Airlines in 2019", 14 December, available online at: https://www.airlines.iata.org/news/iata-forecasts-355bn-net-profit-for-airlines-in-2019
IATA (2018b). Airport Ownership and Regulation, available online at: www.iata.org/policy/infrastructure/Documents/Airport-ownership-regulation-booklet.pdf
IATA (2019a). "Slowing Demand and Rising Costs Squeeze Airline Profits", 2 June, available online at: https://www.iata.org/pressroom/pr/Pages/2019-06-02-01.aspx
IATA (2019b). *Airport Development Reference Manual* (eleventh edition). Montreal: IATA.
IATA (n d). "Airport Infrastructure Investment: Best Practice Consultation", available online at: www.iata.org/whatwedo/ops-infra/airport-infrastructure/Documents/Airport-InfrastructureInvestment-Best-Practice-Consultation.pdf
ICAO (2011). "Safety: Air Navigation Bureau: Making an ICAO Standard", available online at: www.icao.int/safety/airnavigation/Pages/standard.aspx
ICAO (2012a). "Worldwide Air Transport Conference, Sixth Meeting: Slot Allocation", available online at: www.icao.int/Meetings/atconf6/Documents/WorkingPapers/ATConf6-wp011_en.pdf
ICAO (2012b). "Policies on Charges for Airports and Air Navigation Service", ICAO Doc 9082 (ninth edition).
ICAO (2014). "Best Practice in Economic Regulation: Lessons from the UK", symposium presentation, available online at: www.icao.int/Meetings/GACS/Documents/Speaker%20Presentations/Day%204/2-Stephen.Gifford.UK%20CAA.pdf
ICAO (2016). International Standards and Recommended Practices, Annex 14 to the Convention on International Civil Aviation: Aerodromes, Volume I: Aerodrome Design and Operations (seventh edition), available online at: www.icao.int/APAC/Meetings/2016%20ICAOPIS/3%20ICAO%20Annex%2014%20Standards%20and%20Aerodrome%20Certification.pdf
ICAO (2017). International Standards and Recommended Practices, Annex 9 to the Convention on International Civil Aviation: Aerodromes, Volume I: Aerodrome Design and Operations (fifteenth edition), available online at: www.icao.int/WACAF/Documents/Meetings/2018/FAL-IMPLEMENTATION/an09_cons.pdf
ICAO (n d). "The history of ICAO and the Chicago Convention", available online at: www.icao.int/about-icao/History/Pages/default.aspx
ICF (2018). "Insights into the logics of airfares", ACI Europe.
ICSID (2019). "Written Procedure: ICSID Convention Arbitration", available online at: https://icsid.worldbank.org/en/Pages/process/Written-Procedure-Convention-Arbitration.aspx
Inadomi, H.M. (2010). *Independent Power Projects in Developing Countries: Legal investment protection and consequences for development*. Alphen aan den Rijn, Netherlands: Walter Kluwer Law & Business.
Institute for Democracy and Economic Affairs (2018). "Economic Benefits of ASEAN Single Aviation Market", Policy Paper no. 56, available online at: www.ideas.org.my/policy-paper-no-56-economic-benefits-of-asean-single-aviation-market/
Irwin, T. (2007). *Government Guarantees: Allocating and Valuing Risk in Privately Financed Infrastructure Projects*. Washington DC: The World Bank.
Irwin, T. (2008). "Controlling spending commitments in PPPs", in G. Schwartz, A. Corbacho and K. Funke (eds), *Public Investment and Public-Private Partnerships: Addressing Infrastructure Challenges and Managing Fiscal Risks*. Basingstoke: Palgrave Macmillan.
Isaacson, W. (2007). *Einstein: His Life and Universe*. New York: Simon & Schuster.

Kaplan Kirsch & Rockwell (2017). "P3 Airport Projects: An introduction for airport lawyers", available online at: www.kaplankirsch.com/News-Publications/Publications/107702/Evaluating-Airport-P3-Projects-An-Introduction-for-Airport-Lawyers

Klein, M. (1996). "Risk, Taxpayers and the Role of Government in Project Finance", World Bank Policy Research Working Paper 1688, available online at: http://documents.worldbank.org/curated/en/528681468739139955/Risk-taxpayers-and-the-role-of-government-in-project-finance

Kohon, J., Polo, C. and Ricover, A. (2011). *La Infraestructura en el Desarrollo Integral de América Latina, Diagnóstico estratégico y propuestas para una agenda prioritaria. Transporte. IdeAL 2011.* Bogotá: CAF-Banco de Desarrollo de América Latina.

Liet-Veaux, G. and Thuillier, A. (1991). *Droit de la Construction* (tenth edition). Paris: Edité par Litec.

Mackie, K.J., Miles, D. and Marsh, W. (1995). *Commercial Dispute Resolution: An ADR Practice Guide.* London: LexisNexis UK.

National Audit Office, UK (2001). "Getting value for money from procurement", archived online at: https://webarchive.nationalarchives.gov.uk/20130802195149/https://www.nao.org.uk/report/getting-value-for-money-from-procurement-how-auditors-can-help/.

National Audit Office, UK (March 2007). "Improving the PFI Tendering Process", available online at: www.nao.org.uk/report/improving-the-pfi-tendering-process/

Prebble, R. (1985). Parliamentary Debates. House of Representatives, Volume 466, New Zealand Parliament.

Ricover, A. and Delmon, J. (2016). "Mythbusters: Getting Airport PPPs off the Ground", *IFC Handshake*, October, available online at: http://blogs.worldbank.org/ppps/mythbusters-getting-airport-ppps-ground

Ricover, A., Serebrisky, T. and Suárez-Alemán, A. (2018). Comparación de costos aeroportuarios y regulaciones laborales. Washington DC: Banco Interamericano de Desarrollo.

Rodriguez, J.P. et al. (2009). The *Geography of Transport Systems.* Hofstra University, Department of Global Studies & Geography.

Schwartz, J., Andres, L.A. and Dragoiu, G. (2009). "Crisis in Latin America Infrastructure Investment, Employment and the Expectations of Stimulus", World Bank Policy Research Working Paper No. 5009.

Scott, D.K. (2011). "Airport Privatization: Magic Formula or Pandora's Box?", Presentation delivered at Airport Evolution: Latin America Conference, Brazil, May.

Scriven, J., Pritchard, J. and Delmon, J. (eds) (1999). *A Contractual Guide to Major Construction Projects.* London: Sweet & Maxwell.

Serebrisky, T. (2012). *Airport Economics in Latin America and the Caribbean: Benchmarking, Regulation, and Pricing.* Washington DC: World Bank.

Serebrisky, T., Schwartz, J., Pachón, M.C. and Ricover, A. (2011). "Making a Small Market Thrive: Recommendations for efficiency gains in the Latin American air cargo market", World Bank Working Paper, available online at: http://documents.worldbank.org/curated/en/209791468277500221/Making-a-small-market-thrive-recommendations-for-efficiency-gains-in-the-Latin-American-air-cargo-market

Spira (2008). *PPP Case Study: Long-Term Term Concession for Bourgas and Varna Airports.* Bulgaria: U.S. Agency for International Development, May 25.

Straub, J.L., Laffont, J.J. and Guasch, S. (2005). "Infrastructure Concessions in Latin America: Government-led renegotiations", World Bank Policy Research Working Paper, available online at: http://documents.worldbank.org/curated/en/306081468300325193/Infrastructure-concessions-in-Latin-America-government-led-renegotiations

Straub, S. (2008). "Infrastructure and Growth in Developing Countries: Recent Advances and Research Challenges", World Bank Policy Research Working Paper No. 4460. Washington DC: World Bank.

Sutton, D. St J., Kendall, J. and Gill, J. (eds) (1998). *Russell on Arbitration*. London: Sweet & Maxwell.

Tremolet, P. Shukla and C. Venton (2004). "Contracting Out Utility Regulatory Functions, Final Report", January. Washington DC: World Bank.

United Nations Industrial Development Organization (1996). UNIDO Guidelines for Infrastructure Development through Build-Operate-Transfer, available online at: www.unido.org/guidelines-infrastructure-development-through-build-operate-transfer-bot-projects

US Department of Transport/FAA (n d). Advisory Circular on Airport Master Plans AC 150/5070-6B, available online at: www.faa.gov/regulations_policies/advisory_circulars/index.cfm/go/document.information/documentID/22329

Vinter, G. and Price, G. (2006). *Project Finance: A Legal Guide*. London: Thomson/Sweet & Maxwell.

Wallace, I.N.D. (1986). *Construction Contracts: Principles and Policies in Tort and Contract*. London: Sweet & Maxwell.

Wood, P. (2007). *Project Finance, Securitisations and Subordinated Debt* (second edition). London: Sweet & Maxwell.

Zakrzewski, D. (2006). "Airport Privatization: Success or Failure? The Airport Performance Scorecard – A theoretical assessment tool", *Aerlines Magazine* (e-zine edition) 34, November 1.

Index

Note: Page numbers in *italics* indicate figures; page numbers in **bold** indicate tables. The letter 'n' after a page number indicates a note.

accountability 4–5, 238
Addis Ababa Airport **31**, **63**
advisory services 159–61, 220
AENA (Spain) 38, **96**, 122, 137
aeronautical revenues: overview of 26, 26–7, 29–32; and measuring airports' performance 105; regulation of 27, 45, 46–7, 91, 92–3, 93–4; *see also* commercial revenues; non-aeronautical revenues; revenues
Aéroports de Paris 21, 45, 115, 129
Africa: airline profitability 20; airline revenue 23, **31**; examples of concessions 117, **120**, **121**, 131–3; legacy carriers 21; passenger traffic share by airport ownership *132*; and social impact of airports 203
agreements for PPPs: contractual mechanisms 209–18; key issues 191–6; risk allocation 197–209
Air France-KLM 21, 23
air service agreements 77–86
air traffic movements (ATM): demand for 51–2; and operating expenses 38, *39*, 41, *41*, *110*; and total revenues 34, *35*; *see also* passenger traffic; traffic
Airbus 51, 52
airfares: and airline profitability 20; and ancillary services 23; and BASAs 77, 80–1; and introduction of new routes 85; and Open Skies Agreements 81; and overall fees/charges 27; and pressure to lower 11
Airport Development Reference Manual 73, 73–5, 104
Airport Handling Manual 73

airport operators: and commercial services 26; implementation plans 222–3; importance of reputable operators 137; and investors as 'operators' 182; and prequalification process 181–2, 182–3; and public face of services 213; and regulation of fees/charges 27; and selecting PPP partners 135; as strategic investor 12
Airport Privatization Pilot Program (USA) 5
airside areas (airports): and alternative scope of airport PPP *112*; and compliance with regulations 43; planning of facilities 7, 44; and scope of an airport PPP 111, 168; traffic forecasting 161
American Airlines 21, 21n10, 23
ancillary revenues 23, 26, 27–8; *see also* aeronautical revenues; commercial revenues; non-aeronautical revenues; revenues
Antananarivo Ivato Airport 43, **63**, 103, **121**, *133*
arbitration (dispute resolution) 238–9
Argentina 21, **120**, **128**, 129, 195, 226; and Buenos Aires airports 17, 28, 28n4, 48, 58, **63**, **120**, **128**, 222
Armstrong, Neil 219
Arturo Merino Benítez Airport *see* Santiago de Chile Airport
ASEAN Multilateral Agreement on Air Services 82–3
Asia: airline revenue 23, **31**; legacy carriers 21; and management of airports 107

Asia Pacific (APAC): airline operating expenses 38, 38, 39; airline profitability 20; examples of PPP schemes 131, **131**; non-aeronautical revenues 33, 34; passenger traffic share by airport ownership 130; public sector (SOE) airports 129, 130

assets, condition of 210

Association of Southeast Asian Nations (ASEAN) 82–3

Athens Eleftherios Venizelos Airport *see* Eleftherios Venizelos Airport (Athens)

ATM (air traffic movements) *see* air traffic movements (ATM)

Auckland Airport 102, 113, **116**, **131**

Australia: Brisbane Airport 38, 51, **63**; charter airline market 22; examples of share sales **116**, 129, **131**; policy decision on PPPs 101; privatisation of airports 5; and renegotiation process 232–3; SOE airports 107; Sydney Airport 5, 29, 87–8, **96**, 113

Austria (Vienna International Airport) 30, **96**, **116**, **125**

Avianca Airlines 21, 22

awarding a project 156, 187–8, **188**

Azul Linhas Aéreas 17–18, 21, 23

baggage fees 23

baggage handling systems 28, 37, 201, 224

Baggage Reference Manual 73

Bangkok Suvarnabhumi International Airport **31**, 58, 200–1

bankability of projects 140, 146–7, 153, 155, 188

Belgium: Brussels Airlines 21; Brussels Airport 50, **96**, **116**, **125**; examples of share sales **116**, **125**

Belgrade Nikola Tesla Airport 30, **121**, **126**

Ben Gurion Airport (Tel Aviv) 17, 18, 36, **63**

benchmarking 36, 179, 183, 187

Benito Juarez Airport (Mexico City) 52, **63**

Berlin Brandenburg Airport 224

Bermuda 134

best and final offer (BAFO) 188

Bhutan 49

bid (tender) process: overview of 175–6; award/financial close 187–9; bidding 183–7; competitive processes 176–81; and good project preparation 155, 176; and land acquisition 12; pre-qualification process 181–3; transaction procurement *177*

bilateral air service agreements (BASAs) 77–81

boarding bridges 9, 18, 26, 27, 30, **32**

boarding fees 26, 27

Boeing 51, 52

Bolivia 84, 147, 198

Borges, Jorge Luis 190

Braude, Jacob M. 25

Brazil: airlines' ancillary services 23; Azul Linhas Aéreas 17–18, 21, 23; Campinas Airport 18; examples of concessions **120**, **128**, 129, 129n13; and passenger traffic 48; São Paulo Guarulhos Airport 30, 49, **63**, **96**, **120**, **128**

Brisbane Airport 38, 51, **63**

British Airports Authority (BAA) 4, 5, 101, 115

British Airways: ancillary services 23; and mergers 21; slot allocation 87, 88

Brussels Airlines 21

Brussels Airport 50, **96**, **116**, **125**

Bucharest Henri Coandă Airport (Otopeni) 3, 45, 147, 198

Budapest airports 48, 90, **96**, **116**, **125**

Buenos Aires airports 17, 28, 28n4, 48, 58, **63**, **120**, **128**, 222

Build, Operate and Transfer (BOT) model 56, 115–17

Build, Own, Operate and Transfer (BOOT) model 117

bundling 9, 10, 163–4

business journals 170

business plans 62, 148–9

Cairo International Airport 5, 104, **122**

Cambodia 49, 82, **120**, **131**, **131**

Campinas Airport (Viracopos) 18

Canada: airport planning 59; policy decision on PPPs 102; regional airports 11

Cancún Airport 47, **96**, **121**, **128**

capacity of airlines: and BASAs 77; and expansion of facilities 44; and MASAs 82, 83, 84; and slot allocation 86–7

Cape Verde 58

capital expenditures (capex): overview of 42–5; and cost-based regulation 91; as key issue of PPPs 12; and pre-feasibility study 162; and process of

master planning 65, 65; and single-till regulation 97; and state budgets 103; *see also* operating expenditures (opex)
capitalisation of airports **116**
car-parking facilities 28, 36, 111, 192
cargo operations: examples of airports **63**; exercise of freedoms 79, 82, 84; myths around 8–9; and profitability 19, 20; and traffic forecasting 60, 161
Caribbean: concessions 129; passenger traffic 48, 50, *127*
CARICOM Multilateral Air Service Agreement 83–4
Centre for Aviation (CAPA) 18–19
Changi Airport (Singapore) 18, 22, *38*, 58, **63**, 107
charges basket approach 94–5
Charles de Gaulle International Airport (Paris) 30, **31**, 51, **63**, **116**
charter airline market 22–3, 49
check-in services 9, 17, 18, 26, **76**
Chicago Convention (ICAO) 68–9, 70–2, 77, 78, 79, 103n4
Chicago Midway Airport 178
Chile: adoption of BOT model 117; airline revenue 46; economic regulation of airports 99; examples of concessions **120**, **128**, 129, 164; and good PPP project preparation 164; granting of ninth freedom 79; legacy carriers 21; and management of airports 104; passenger traffic 10; privatisation of airports 6; and renegotiation process 232; signing of MASA 82; *see also* Santiago de Chile Airport
Cincinnati/ Northern Kentucky International Airport 50
Ciudad Real Airport 11, 59
Civil Aviation Authority (UK): and economic regulation of airports 98–9, 99n38; and expansion of airport facilities 44–5
cleaning services 27, 36, 161
Colombia: El Dorado Airport 22, 52–3, **63**, 103, 111; examples of concessions **120**, **128**, 129; legacy carriers 21; and renegotiation process 234; signing of MASA 84; and transport modes 50; and unsolicited proposals 180, 181
commercial revenues: overview of 26, 28–9; and management of airports 104, 107; revenues per passenger by airport ownership 107, *108*; single vs dual till

91–2, 93, 94; value of feasibility study 165; and vertical integration at airports 23–4; *see also* aeronautical revenues; non-aeronautical revenues; revenues
commercial viability of projects 47, 155
competition: and airline profitability 19, 20; myths around 3, 9; and procurement process 176–81; and regulation of fees/ charges 27
completion date 210
completion risk 149, 197, 199–200; *see also* risk
completion time 212
compliance monitor 226
compliance standards 43, 75, 104, 162, 203; *see also* regulation of air transport
Concepción Airport 6
concession fees 115, 117, 196, 211; *see also* fees
concession models: and air transport regulation 98; and alternative scope of airport PPP *114*; and bidding process 155; BOT model 56, 115–17; contractual structure of *118*; factors minimising government contributions 56; geographical examples of 5, **120–1**, **125–6**, **128**, **129**, **131**, **133**; and management of airports 104; and measuring airports' performance 106, 107, *108*, *109*, *110*; and passenger traffic *122*, *124*, *127*, *130*, *132*; payment structure of *119*; PPP schemes for landside areas 112; and securing of financing 184; types of PPP contracts by region *123*
Condor 22, 23
conflict management 196, 233, 235–40
congestion pricing 92, 92–3
connecting traffic 8, 50, 60–1; *see also* passenger traffic; traffic
construction and operation activities 210, 216, 224–5
construction contractors 135, 182, 211–13, 214
consultation processes 65–6, 169, 237
contracts: and level of service (LoS) 75; and PPP agreements 194–5, 209–18; PPP contracts by region *123*; service contracts 35, 35n8, 36, 37, 38; signing of 188; structure of concession model *118*; *see also* management contracts
Copa Airlines 21, 49, 66
Copenhagen airports 18, 89, **96**, **116**, **125**

corporate financing 141, 142
corporatisation 4–5
cost-based regulation 89–90, 91; *see also* economic regulation of airports
Costa Rica: examples of concessions **120**, **128**; Juan Santamaría Airport 22, 48, **120**, **128**, 234; legacy carriers 22; Liberia Airport 36, 48, 112; and renegotiation process 234
CPH Go Airport (Copenhagen) 18
credit enhancement 144, 217–18
currency risk 196, 202; *see also* risk
CUTE (Common Use Terminal Equipment) 28, 28n2

debt: contributions 143–4; and credit enhancement 144, 217–18; debt competition 228; subordinated debt xviii, 144, 146; tenor of 55, 141, 166, 201, 228
debt service cover ratio 149, 214
debt to equity ratio 149, 214
Decision on Integration of Air Transport of the Andean Community 84
default, events of 206, 214
defects liability 213, 224
Delta Air Lines 21, 23, 50
demand risk 50, 51, 62, 202–3; *see also* risk
Denver International Airport 5, 58, 224
design of airports: and expansion of facilities 44; and PPP agreements 192–3; regulation of 14
designation of airlines 77, 81, 84
development fees 32, 33
dispute resolution 196, 233, 235–40
divestiture/equity sale model 112, 113–15; *see also* share sale models
Dominican Republic: examples of concessions **120**, **128**; and open skies agreements 50; Punta Cana International Airport 5, 33, 58
dual till regulation 46–7, 91, 91–2, 92, 93, 94, **96**; *see also* till regulation
Dubai airports 36, 51, 58, **63**
Düsseldorf Airport 38, **63**

earnings before interest, taxes, depreciation, and amortisation (EBITDA) 39, *40*, 41, *42*
easyJet 16, 17, 18, 21, 23
economic context: and feasibility studies 166; and passenger traffic 48, 61; and project financing 147–8

economic development: and airport planning 69–70; and role of airports viii, 11; and traffic 50, 61
economic internal rate of return (EIRR) 147
economic regulation of airports 88–99; *see also* regulation of air transport
Ecuador: and bidding process 178; examples of concessions **120**, **128**, 129; legacy carriers 21; Mariscal Sucre International Airport 45, 60, **96**, **120**, **128**, 147, 178, 197–8, 231; and renegotiation process 231; signing of MASA 84
Egypt: Cairo International Airport 5, 104, **122**; and good PPP project preparation 161; and management of airports 104; passenger traffic 48; privatisation of airports 5
Einstein, Albert 15
El Dorado Airport (Bogotá) 22, 52–3, **63**, 103, 111
El Loa Airport 46, **120**, **128**
El Palomar Airport 17, **63**, **120**
Eleftherios Venizelos Airport (Athens): airline revenue 28, 30, 33; and airport planning 60; and capital subsidies 12; distribution of opex for airport groups *38*; as international terminal airport **63**; and profitability of airports 58; till regulation **96**
employees (airport) 42, 194, 222
Enhancement and Financing Services (IATA) 72, 75–6
Enterprise Team (ET) 221–2
environmental issues: and ICAO standards 70; and PPP agreements 195; and traffic forecasting 61
environmental risks 52, 203–4
Equator Principles 203–4
equity contributions 143
equity, selling down 229–30
Ethiopia: Addis Ababa Airport **31**, **63**; legacy carriers 21
Europe: aeronautical revenues **31**; airline operating expenses *38*, 38–39, *39*, *40*; airline profitability 20; examples of concessions **120**, **121**, **122**, **125**, **126**, **126**; examples of share sales 113, **116**, **122**, **125**, **126**, **126**; legacy carriers 21; non-aeronautical revenues *34*; passenger traffic share by airport ownership *124*; SOE airports 107, **122**; *see also* names of countries

Index   251

European Union: ECAA Agreement 85–6; and PPP agreements 192
evaluation: and dispute resolution 236–7; process for proposals 154
exclusivity 193, 210
expansion/planning of facilities 44–5, 52–3, 71, 103
expropriation of property 147, 198, 235

FAA (Federal Aviation Administration) 5, 44, 64
facilities: aircraft parking 18, 44, 47, 193; car parking 28, 36, 111, 192; planning/expansion of 44–5, 52–3, 71, 103
feasibility studies: overview of 164–9; and good project preparation 154–5; pre-feasibility assessment 154, *156*, *157*, 161–3
Federal Aviation Administration (USA) 5, 44, 64
fees: aircraft-related 29–32, 76, 111, 191; and airport management 105; concession-related 115, 117, 196, 211; passenger-related 18, 27, 32–3; regulation of 94–5; *see also* revenues
financing of PPPs: overview of 139–42; and bidding process 184, 186; and credit enhancement 144, 217–18; and financial advisers 160; financial close *156*, 188–9, 227; and financial intermediation 150–1; and financial viability 149, 155, 201; and financing risk 201–2; and good project preparation 153; and government financial mechanisms 53–6; and implementation of projects 226–30; and intercreditor arrangements 215; and lending agreements 213–15; myths around 4; and PPP agreements 196; and pre-feasibility/feasibility studies 162, 166; and project financing xvii, 141–2, 145–9; and refinancing 149–50, 228–9; and shareholder/sponsor support 146, 218; sources of 143–4; types of borrowing 142–3; *see also* investment
financing risk 201–2; *see also* risk
Flybondi 17
force majeure 204–6
forecasting *see* traffic: forecasting
foreign exchange risk 56, 150, 196; *see also* risk
France: Aéroports de Paris 21, 45, 115, 129; Charles de Gaulle International Airport 30, **31**, 51, **63**, **116**; examples of share sales **116**, **125**; and *force majeure* 204; legacy carriers 21, 23; Lyon-Saint Exupéry Airport 17, 18
Frankfurt Airport: airline revenue 45; as airport hub **63**; and demand for ATMs 51; distribution of opex for airport groups 38; and economic regulation of airports 89, 89n31, **96**; landing charges structures **31**; and management of Cairo Airport 104; noise-related issues 30, 52, 149
freedom rights 77–9, 82, 84
fuel costs 10, 19, 165
full business case *see* feasibility studies
full service carriers *see* legacy carriers

Gatwick Airport (London): economic regulation of 95, **96**, 99, 99n38; as example of share sale model 113, **116**, **126**; and leisure-oriented traffic **63**; privatisation process 4; slot allocation 87
geographical distribution of PPPs 4–5, 117–33
Georgia (Tbilisi Airport) 50, **121**, **125**, **126**
Germany: Berlin Brandenburg Airport 224; charter airline market 22; Düsseldorf Airport 38, **63**; examples of share sales 113, **116**, **125**; Munich International Airport 50, 58, 129, 135; *see also* Frankfurt Airport
government support to PPPs: factors influencing revenue potential 165; factors minimising contributions 56; and financing for projects 142, 150–1; and government interference 211; and guarantees for public sector bodies 210; legal/regulatory context 191; policy motivations for PPPs 101–2; and use of appropriate instruments 53–6
Greece: airport planning 60; examples of concessions **121**, **125**; *see also* Eleftherios Venizelos Airport (Athens)
Ground Operations Manual 73
Grupo TACA 21, 22
Guarulhos Airport (São Paulo) 30, 49, **63**, **96**, **120**, **128**

Haneda Airport 59–60, 107
Heathrow Airport (London): as airport hub 58, **63**; baggage handling system 201, 224; and consolidation of airlines

22; economic regulation of 95, 96, **96**, 99, 99n38; as example of share sale model 113, **116, 126**; noise reduction efforts 30; operating expenses 36, 37, 38; privatisation process 4; proposed expansion of 52, 66, 149; slot allocation 87, 88
Honduras 49, **121, 128**
Hong Kong International Airport **31**, 38, **63**, 107, 224
hubs: and airport planning 58, **63**; and demand for ATMs 51; myths around 8–9
Hungary: Budapest airports 48, 90, **96, 116, 125**; and political context for concessions 147
hybrid till regulation 46, **96**; *see also* till regulation

IAG (International Airlines Group) 21, 88
IATA (International Air Transport Association) *see* International Air Transport Association (IATA)
Iberia Airlines (Spain) 21, 50
ICAO (International Civil Aviation Organization) *see* International Civil Aviation Organization (ICAO)
implementation of projects: overview of 219–20; and dispute resolution 235–40; and expiry/handover 240; and financing 226–30; and good project preparation 155; monitoring of 220–6; as phase of PPP process *156*; and renegotiation processes 230–4; and step-in 234–5
India: examples of concessions 99, 107, 117, **121**, 131, **131**; and number of airlines 21
Indonesia 11, 21, 82, 180
industry journals 170
information memorandum *156*, 164, 170, 172–4, 174n12, 183
infrastructure: and airport planning 59–60, 62; and private sector financing of 4n2, 103; and sources of revenue 33
Instrument Landing System (ILS) 32, 32n7
insurance coverage 215–17
interest rates 4n2, 141, 201
International Air Transport Association (IATA): and air transport regulation 72–6; and airline profitability 19–20; and airports' compliance with standards 104; and slot allocation 86, 86–7

International Airlines Group (IAG) 21, 88
International Civil Aviation Organization (ICAO): and air transport regulation 68–72, 94, 97; and airport planning 64–5; and compliance standards 43; and slot allocation 86, 87
investment: and air service agreements 82; and airport planning viii–ix; creating investor appetite 183, 187; and economic regulation of airports 93–5; factors minimising government contributions 56; and good project preparation 153, 155; importance of local investors 135, 182; investors as 'operators' 182; and PPP agreements 192–3; and selling down equity 229–30; *see also* financing of PPPs
Israel: Tel Aviv Ben Gurion Airport 17, 18, 36, **63**; Timna Ramon International Airport 59
Istanbul New Airport 126, 126n10
Italy: and distribution of opex for airport groups 38; examples of share sales **116, 125**; Malpensa Milano Airport 17, 18, **116, 125**; Rome-Fiumicino International Airport **31, 96, 116, 125**

Jamaica: Montego Bay Airport 142; Norman Manley International Airport 33, 43, **121, 128**, 137
Japan: air transport regulation 99; examples of concessions **121**, 131, **131**; Haneda Airport 59–60, 107; legacy carriers 21; Narita Airport 17, 22, 59–60, **63**
JetSmart 17
JFK Airport 5, 30, **31**, 112
Jomo Kenyatta International Airport (Nairobi) 60
Jordan 5, **121, 133**
Jorge Chávez Airport (Lima): as example of concession **121**; and land acquisition 12; and profitability of airports 58; as regional hub **68**; and selection of PPP partner 135, 137; sources of revenue 28, 30; upgrading of facilities 103
journals, industry/business 170
Juan Santamaría Airport (San José) 22, 48, **120, 128**, 234

Kathmandu Tribhuvan International Airport 28, 49
Kelleher, Herbert 16
Kenya: Jomo Kenyatta International Airport 60; Kenya Airways 21

King, Rollin 16
KLM Airline 21, 23
know-how transfer 185, 209, 210–11
Kuala Lumpur International Airport 18, **116**, **131**
Kyiv Zhuliany International Airport 17, **63**

labour contracts 16
labour costs 19, 35, 36, 37, *37*, 38, 39, 107
labour relations 42, 194, 199
land acquisition/use: and airport master plan 64; and bankability of projects 148; as key issue of PPPs 12; legal/regulatory contexts 199; and PPP preparation 167; and social impact of airports 61, 70, 203
landing fees 29–30, **31**, 76, 111, 191
landside areas (airports): and alternative scope of airport PPP *112*; compliance with regulations 43; cross-subsidisation of aeronautical services 93; planning of facilities 44–5; and scope of an airport PPP 111, 112, 168, 192; and traffic forecasting 161–2
LATAM Airlines Group 21, 23
Latin America: airline operating expenses 38, *38*–9, 39, *40*; airline profitability 20, *20*; airline revenue 23, **31**, 33, *34*; examples of concessions 5, 107, 117, **120**, **121**, **128**, 129; legacy carriers 21, 22; and renegotiation processes 230; share of passenger traffic *127*; signing of MASA 84; upgrading of airport facilities 103; *see also* names of countries
legacy carriers: overview of 18–22; and air service agreements 85; and basic economy fares 23;
legal advisory services 160
legal context for agreements 191, 197–8, 198–9, 204–5
legal viability of projects 155, 167, 186
lending agreements 213–15
levels of service (LoS) (IATA) 74–5, **76**, 104, *104*–5, 193, 210
liability for defects 213, 224
liability insurance 216
liberalisation of aviation market 79–81, 83, 165
Liberia Airport 36, 48, 112
litigation 240
load factors 9, 17, 20, 45, 45n16, 52
loan life cover ratio 149, 214
loans, subordinated 144

London City Airport 87, **116**, **126**
Los Angeles (LAX) Airport **31**, 38
loss insurance 216
low-cost carriers (LCCs): overview of 16–18; and air service agreements 83, 85; and alternative airport roles **63**; ancillary services 23; and charter airline market 22; in Europe 21; vs legacy carriers 18–19; and load factors 52; myths around 9;
Lufthansa 21, 22–3, 23, 24, 50
Lyon-Saint Exupéry Airport 17, 18

Macedonia 85, **121**, **125**; and Skopje Alexander the Great Airport 30, **121**, **125**, 126
Madagascar: Antananarivo Ivato Airport 43, **63**, 103, **121**, **133**; Nosy Be Fascene Airport 43–4, 58, 103, **121**, **133**; opposition to airport expansion 52; privatisation of airports 6
maintenance: costs 35–6, 37, *37*, 38, 43–4; maintenance, repair and overhaul (MRO) facilities 9–10; operation and maintenance agreement 213; regime 210
Maiquetía Airport 59
Malaysia: airports' operating expenses 38; examples of share sales **116**, 129, **131**; Kuala Lumpur International Airport 18, **116**, **131**; signing of MASA 82
Male International Airport (Maldives) 3, 48, 72, 147
Malpensa Milano Airport 17, 18, **116**, **125**
management contracts: overview of 117; and advisory services 159; and bidding process 155, 176; and bundling 163; and commercial activities 165; duration of 113; geographical examples of 5, **122**, *123*, **125**; and ownership 113; and risk transfer 113; and size of airport 10; *see also* contracts
management of airports: improvements in 104–5; and project team 220
Manchester Airport 87, 95
Mariscal Sucre Airport (Quito) 45, 60, **96**, **120**, **128**, 147, 178, 197–8, 231
market sounding 156, 169–70, 186
Marx, Groucho 67, 100
master plans (airports) 44, 62–5, *156*, 165, 193
Maximum Takeoff Weight (MTOW) 29–30, 29n5, 30, **31**, **32**
measurement of PPP success 105–11

254 Index

mediation processes 233, 237–8
mergers, airlines 21–2, 21n10
Mexico: and airport operators 135, 137; Benito Juarez Airport 52, **63**; and bidding process 182; and bundling 164; Cancún Airport 47, **96**, **121**, **128**; and economic regulation of airports 95, 99; environmental concerns around new airport 149; examples of concessions **121**, **128**, 129
mezzanine financing xvi, 144; *see also* financing of PPPs
Middle East: airline profitability 20; airline revenue 23; assigned seat fees 23; examples of concessions **121**, 131, 133, **133**; passenger traffic share by airport ownership *132*
minimum technical requirements (MTRs) 75, 105, *156*, 192, 193, 225
Mirabel Airport 11, 59
Molière 175
monitoring of projects 225–6
Montego Bay Airport 142
Mozambique 49
multilateral air service agreements (MASAs) 81–6
Munich International Airport 50, 58, 129, 135

Nairobi Jomo Kenyatta International Airport 60
Narita Airport (Tokyo) 17, 22, 59–60, **63**
nationalisation of airports 198
natural disasters 48, 61, 69, 205
negotiation (dispute resolution) 237
Netherlands: distribution of opex for airport groups *38*; examples of share sales 113, **125**; *see also* Schiphol Airport (Amsterdam)
network carriers *see* legacy carriers
New York Convention 239, 239n19
New Zealand: air transport regulation **96**; Auckland/Wellington Airports 102, 113, **116**, **131**; corporatisation process 5; examples of share sales **116**, 129, **131**; policy decision on PPPs 102; signing of MASA 82; SOE airports 107
Nikola Tesla Airport (Belgrade) 30, **121**, **126**
noise: charges 27, **31**; and impact of 149; listed in information memorandum 173; and reduction/regulation of 30, 52, 70, 193

non-aeronautical revenues: ancillary services 23, 26, 27–8; and cost drivers 45–7; myths around 8; regulation of 91, **92**; variations in 33–4, *35*; *see also* aeronautical revenues; commercial revenues; revenues
Norman Manley International Airport (Kingston) 33, 43, **121**, **128**, 137
North America: airline operating expenses 38, 38–39, *39*, *40*; airline profitability 20, *20*; airline revenue **31**, *34*; legacy carriers 23; SOE airports 107; *see also* United States (US)
Nosy Be Fascene Airport 43–4, 58, 103, **121**, **133**

open skies agreements 50, 81, 82, 83
operating expenditures (opex): overview of 34–42; per ATM by airport ownership *110*; per passenger by airport ownership *107*, *109*; and till regulation 97; *see also* capital expenditures (capex)
operation and maintenance agreements 213; *see also* maintenance
operation manuals 223
operation risk 201; *see also* risk
operational readiness and airport transfer (ORAT) 223–4
operators *see* airport operators
O.R. Tambo Airport (Johannesburg) 30, **63**, 87, **96**
origin and destination (O&D) traffic 8, 48–50, 61, 165; *see also* passenger traffic; traffic
OSITRAN (Peru) 98, 225–6
ownership in PPPs 101, 113

Pacific Airports Group 163
Panama: Copa Airlines 21, 49, 66; Tocumen International Airport 49, 58, 66, 129
parking facilities: for aircraft 18, 44, 47, 193; for cars 28, 36, 111, 192
parking fees 30, **32**
Paro International Airport 49
passenger service charges 18, 27, 32–3
passenger traffic: and airport ownership *122*, *124*, *127*, *130*, *132*; connecting traffic 8, 50, 60–1; global figures 15–16, 117; myths around 7, 10, 47; O&D traffic 8, 48–50, 61, 165; and transport modes 50; *see also* air traffic movements (ATM); traffic

Index   255

performance of a project 210
performance risk 200–1, 212, 213; *see also* risk
personnel (airport staff) 42, 194, 222
Peru: air transport regulation 98, 225–6; examples of concessions **121**, **128**, 129; legacy carriers 21; privatisation of regional airports 6; and renegotiation process 234; signing of MASA 84; *see also* Jorge Chávez Airport (Lima)
Philippines 21, 82, **121**, 131, **131**
planning of airports: and associated infrastructure 59–60; consultation process 65–6; master planning 62–5; strategic considerations 57–9; and traffic forecasting 60–2
political context: and economic regulation of airports 89; and passenger traffic 48, 61; political risks for PPPs 197–8, 206; and project financing 147–8
Porlamar International Airport 3, 198
Portugal: and distribution of opex for airport groups 38; examples of share sales 113, **116**, **125**; passenger traffic 50, 122
Prague Vaclav Havel Airport 30
pre-feasibility studies 154, *156*, *157*, 161–3; *see also* feasibility studies
pre-qualification process 181–3
preference shares 144
preparation process for PPPs *see* project preparation
price cap regulation 90–1
price of tickets *see* airfares
private sector: and efficiency 3; and performance of airports 105–6; and PPP financing 4, 4n2, 5–6, 140–1; and renegotiation processes 230, 231; *see also* public private partnerships (PPPs)
procurement process *see* bid (tender) process
profitability of airlines: and airport planning 58; drivers of 19–20; and operating expenses 41–2; *see also* revenues
project development funds (PDF) 55–6
project financing xvii, 141–2, 145–9; *see also* financing of PPPs
project preparation: overview of 152–5; and approval regime 171; and consultation 169; and feasibility study 154–5, 164–9; and market sounding

156, 169–70, 186; and pre-feasibility study 154, *156*, *157*, 161–3; and project selection 155–9
project team 159–61, 220–2
promotion of projects 169–70
proposals, unsolicited 176, 178, 179–81
public private partnerships (PPPs): advantages of 2; geographical distribution of 4–5, 117–33; key issues 12–14; and measuring airports' performance 105–11; myths around 2–11; phases of developing a PPP airport *156*; scope of a PPP 111–12, *114*, 168, 191–3; selecting the correct form of 133–5; selection of partner 135–8; *see also* concession models; management contracts; share sale models
public sector: and efficiency 3; vs private sector financing 4
public sector comparators 134, 168
public sector (SOE) airports: and implementation of PPP agreements 195; and measuring airports' performance 105, *106*, 107, *108*, *109*, *110*, 111; passenger traffic share by airport ownership *122*, *124*, *127*, *129*, *130*, *132*
Pulkovo Airport (St Petersburg) 5, 13, 45, **121**, **126**, 159, 199
Punta Cana International Airport 5, 33, 58

Queen Alia International Airport 5, **121**, **133**
Quito Mariscal Sucre International Airport 45, 60, **96**, **120**, **128**, 147, 178, 197–8, 231

ramp handling services 9, 28, 36, 45, 194, 195
rating agencies 104, 144, 217
real estate 28–9, 51, 56, 62
refinancing of projects 149–50, 228–9; *see also* financing of PPPs
refinancing risk 166, 196, 229; *see also* risk
regional airports: and demand for ATMs 51; examples of concessions **121**, **125**, **128**; and hubs **63**; listed in information memorandum 174; privatisation process 5, 5–6; and regional development 11; regulation of 89
regulation of air transport: and aeronautical revenues 27, 45, 46–7, 91, 92–3, 93–4; and compliance standards

43, 75, 104, 162, 203; and economic regulation 88–99; and high costs 17; as key issue of PPPs 14; market access and air service agreements 77–86; and passenger traffic 49, 50; and PPP agreements 191, 193; and slot allocation 86–8; and traffic forecasting 61; *see also* International Air Transport Association (IATA); International Civil Aviation Organization (ICAO)
renegotiation processes 156, 230–4
repatriation of funds 79, 202
request for proposals (RfP) 183
revenue yield approach 94–5
revenues: and airline profitability 19, 20; and economic regulation of airports 88–99; factors influencing revenue potential 165; as key issue of PPPs 12, 14; and PPP agreements 195–6; total revenues per ATM 34, 35; total revenues per passenger by airport ownership 105, 106, 115; *see also* aeronautical revenues; commercial revenues; non-aeronautical revenues
risk: completion risk 149, 197, 199–200; currency risk 196, 202; demand risk 50, 51, 62, 202–3; financing risk 201–2; and *force majeure* 204–6; foreign exchange risk 56, 150, 196; and insurance coverage 215–17; and operation manuals 223; operation risk 201; performance risk 200–1, 212, 213; political risks for PPPs 197–8; and private vs public interest rates 4n2; refinancing risk 166, 196, 229; risk matrix 167, 223; risk transfer 113, 114; site risk 212; traffic demand risk assessment 50–1
Rogers, Will 139
Romania (Henri Coandă Airport) 3, 45, 147, 198
Rome-Fiumicino International Airport 31, 96, 116, 125
route schedules 77, 80, 81, 83
Russia: examples of share sales 116, 125; privatisation of airports 5; Pulkovo Airport 5, 14, 45, 121, 126, 159, 199
Ryanair 16, 17, 21, 23

safety and security: and airline capital expenses 42–3; decline in standards 1–2; IATA's safety guidelines 73; ICAO standards 69, 70–1, 104; legal/regulatory contexts 191, 232; national security considerations 59; outsourcing of services 36; and performance risk 200; and PPP agreements 105, 192, 198; and renegotiation process 230; screening costs 36; security fees 33; technical requirements 171
Saint-Exupéry, Antoine de 57
salaries *see* labour costs
San José Juan Santamaría Airport 22, 48, 120, 128, 234
San Salvador Airport 22, 129
Santiago de Chile Airport: as example of concession 112, 120, 128; financing of 103, 135, 182; landing fees 30, 31; and renegotiation process 231; sources of revenue 45
São Filipe Airport 58
São Paulo Guarulhos Airport 30, 49, 63, 96, 120, 128
Saudi Arabia 5, 121, 122, 133
Schiphol Airport (Amsterdam): economic regulation of 96; as example of share sale model 116, 125; expansion of facilities 45; and LCCs 18; and mergers 21; operating expenses 37; as regional hub 63
scope of a PPP 111–12, 114, 168, 191–3
seat: capacity 83; cost per available 19; density 16; fees 23
security *see* safety and security
security rights 167, 199, 214–15, 215
sensitivity analysis 166n8
Serbia 85, 121, 126
service contracts 35, 35n8, 36, 37, 38
share sale models: geographical distribution of 113–15, 116, 122, 125, 126, 126, 129, 131; and measuring airports' performance 106, 107, 108, 109, 110; passenger traffic share by ownership 122, 124, 130; type of PPP contract by region 123
share transfers 211
shareholders: shareholders' agreement xviii, 218; and sponsor support 146, 218; *see also* financing of PPPs; investment
Siem Reap Airport 49, 120, 131, 131
Singapore: Changi Airport 18, 22, 38, 58, 63, 107; signing of MASA 82; Singapore Airlines 78
single till regulation 46–7, 91–5, 95–6, 97; *see also* till regulation
site access 193–4
site risk 212; *see also* risk
size: of aircraft 51–2; of airport 33–4, 35, 39–42, 40, 41, 42

Skopje Alexander the Great Airport 30, **121**, **125**, 126
slot allocation 86–8
social impact of airports 61, 70, 203–4
South Africa: and good PPP project preparation 161; O.R. Tambo Airport 30, **63**, 87, **96**; South African Airways 21, 87; and tourist traffic 49
Spain: AENA Spain 38, **96**, 122, 137; Ciudad Real Airport 11, 59; economic regulation of airports **96**; examples of share sales **116**, 122, **126**; Iberia Airlines 21, 50
special purpose vehicle xviii, 141, 142, 146
sponsor support 146, 218
staff (airport personnel) 42, 194, 222
Standards and Recommended Practices (SARPs) 68–9, 69, 104, 162; *see also* regulation of air transport
Stansted Airport (London): economic regulation of 95; as example of share sale model **116**, **126**; as low cost carrier base **63**; privatisation process 4; slot allocation 87
state-owned airports *see* public sector (SOE) airports
step-in rights xviii, 207–8, 208n21, 211, 215, 234–5
subordinated debt xviii, 144, 146
subordinated loans 144
Suvarnabhumi International Airport (Bangkok) **31**, 58, 200–1
Swiss Challenge 178, 180
Switzerland (Zurich Airport) 36, **116**, **126**
Sydney Airport 5, 29, 87–8, **96**, 113

TACA Airlines 21, 22
Tbilisi Airport 50, **121**, **125**, 126
team, project 159–61, 220–2
technical advisory services 160, 171
technical requirements, minimum (MTRs) 75, 105, *156*, 192, 193, 225
technical service agreements 135, 137, 181–2, 211
technical viability of projects 155, 167, 184, 185–6
Tegucigalpa Toncontin Airport 49, **121**, **128**
tender process *see* bid (tender) process
terminal navigation charges 32
termination of a project 206–8, 211, 234–5, 240

Thailand: airline operating expenses 38, 38; signing of MASA 82; Suvarnabhumi International Airport **31**, 58, 200–1
Thomas Cook Group 22, 22–3
ticket prices *see* airfares
till regulation 46–7, 91–6, *97*
Timna Ramon International Airport 59
Tocumen International Airport (Panama) 49, 58, 66, 129
tourism travel: and airport planning 58; charter airline market 22; and consultation process 169; and passenger traffic 47, 48, 49, 50, 61, 165
traffic: factors influencing revenue potential 165; forecasting 50–1, 60–2, 140, 161–2, 165, 170; leisure-oriented **63**; rights 81, 82, 83–4; and volume of 35, 47, 49, 165; *see also* air traffic movements (ATM); passenger traffic
transaction advisers 159–61
transaction procurement *156*
transaction structuring *156*, *158*
transfer of a project 208–9
transport: and links/access 163, 192; and modes of 50
Tribhuvan International Airport (Kathmandu) 28, 49
TUI Group 22, 23
Tunisia 5, **121**, 131, **133**
Turkey **121**, 126, **126**
turnkey construction xviii, 211–12, *212*

Unit Load Devices Regulations 73
United Airlines 21, 23
United Kingdom: air transport regulation 78; economic regulation of airports 95, 95–7, 98; examples of share sales 113, 115, **116**, **126**; and good PPP project preparation 161; policy decision on PPPs 101; privatisation of airports 4–5; and risk allocation 197; slot allocation 87; *see also* Gatwick Airport (London); Heathrow Airport (London); Stansted Airport (London)
United States (USA): air transport regulation 71, 78; charter airline market 22; Chicago Midway Airport 178; Cincinnati/ Northern Kentucky International Airport 50; Delta Air Lines 21, 23, 50; Denver International Airport 5, 58, 224; and *force majeure* 204; JFK Airport 5, 30, **31**, 112; legacy carriers 21; Los Angeles (LAX) Airport

31, 38; passenger traffic 49; privatisation of airports 5; *see also* North America
unsolicited proposals 176, 178, 179–81
Uruguay 79, **121**, **128**
utilities: and financing 142, 150–1; and infrastructure construction 224; as source of operating expenses 35, 36, 37, 38

Vaclav Havel Airport (Prague) 30
value for money (VfM) 133–4, 134–5, 153, 161, 167–8, 187, 231
Venezuela 3, 59, 147, 198, 202
Vienna International Airport 30, **96**, **116**, **125**
VIP lounges 23–4
Voltaire 1

weighted average cost of capital (WACC) 141, 145
Wellington Airport 102, 113, **116**, **131**
Wizz Air 17, 21

Zurich Airport 36, 116, **126**